Springer Proceedings in Physics 138

Springer Proceedings in Physics

Please view available titles in *Springer Proceedings in Physics* on series homepage http://www.springer.com/series/361/

Editors

Man-Wook Han · Jehyun Lee

EKC 2010

Proceedings of the EU-Korea Conference on Science and Technology

Springer

Editors

Man-Wook Han
President Association of Korean Scientists and Engineers in Austria
Verein koreanischer Wissenschaftler und Ingenieure in Österreich,
Simmeringer Platz 1/1/60
Vienna
Austria
Email: han@tuwien.ac.at

Jehyun Lee
Vienna University of Technology
Advanced Magnetism Group
Institute of Solid State Physics
Wiedner Hauptstrasse 8-10
1040 Vienna
Email: jehyun.lee@gmail.com

ISSN 0930-8989 e-ISSN 1867-4941

ISBN: 978-3-642-17912-9 ISBN 978-3-642-17913-6 (eBook)

DOI 10.1007/978-3-642-17913-6

Springer Heidelberg Dordrecht London New York

Library of Congress Control Number: 2011932439

Typeset & Cover Design: Scientific Publishing Services Pvt. Ltd., Chennai, India.

Printed on acid-free paper

Springer is part of Springer Science+Business Media (www.springer.com)

Preface

It is our great pleasure to welcome you to this issue of the Springer Book Series. This issue includes the selected papers written and presented by researchers at the EU Korea Conference on Science and Technology 2010 in Vienna (EKC 2010, http://www.ekc2010.org). After the successful EKC 2008 in Heidelberg, Germany and EKC 2009 in Reading, UK EKC 2010 was held in Vienna Austria. EKC 2010 gave an opportunity to build a network amongst Korean Scientists and Engineers in Korea and Europe. Almost 280 scientists and engineers participated in EKC 2010 to foster the collaboration between Korean and EU as well as to exchange their expertise.

Vienna is a city of art and music boasting in its long tradition of theatre, opera and arts. However Austria is also noted for its science and technology.

From this reason we tried to combine the science and technology with culture. The headline of the EKC 2010 was **(Science+Technology)* Culture**. On the other hand we try to appeal the Austrian science and technology to contribute the development of Korean science and technology.

There was a variety of expert presentations ranging from basic science to advanced technology at the EKC2010, giving plenty of opportunities to exchange information and to become aware of current issues.

We want to express our sincere thanks to organizing committee members, especially, Ki Jun Lee, President of KOFST and Soo Sam Kim, Vice-president of KOFST, Joon Woen Seok, President of VeKNI (Germany), Joung Hwan Lee, President of KSEAUK (UK), Chang Hoon Jun, President of AsCOF (France), volunteers, supports and sponsors. Without their contribution it was not possible to make the event successfully.

Our special thanks go to members of local organizing committee in Austria including Chang Shik Chung, Hey Suck Chung, Jong Mun Park, Chul Woo Hyun, Youn Joo Hyun, Myung Joo Kang, Hong Joon Lim, Eun Young Lee, Gak Hee Lee, Hanna Sheu, Gi Won Chung, Bo Sung Kim and Sylvia Lee who helped a lot for organization of the EKC 2010.

Contents

Determining Potential Areas for Mitigation of Climate Change Risks in Ho-Chi-Minh City

Kiduk Moon

Brandenburg University of Technology Cottbus,
Department of Environmental Planning, Erich-Weinert-Str.1,
D-03044 Cottbus, Germany
moonkidu@tu-cottbus.de

Abstract. The contribution is based on initial research experiences of the development of spatially explicit vulnerability indicators for Impacts of climate change on the high-dense urban area of Ho Chi Minh City using a Matrix of extended Urban Structure Type and Landusemap.

Keywords: Urban Structure Type, Land Use Map, Climate Change Impact, Urban (Re) Development, Green Policy.

1 Introduction

Climate change is altering the traditional context of urban land-use planning and is shaping the priorities of sustainability. The environmental dimension of spatial planning in emerging megacities, such as Ho Chi Minh City (HCMC), has become a strong rationale for coordinating spatially well defined adaptation actions and responses and integrating mitigation policies. The case study is based on initial research experiences of the development of GIS-based vulnerability indicators in the application area of adaptation of a high-risk agglomeration area to the future impacts of Climate Change. Based on the extended 'Urban Structure Type (UST) approach'(HAGGAG & AYAD 2002). The proposed model represents an interpretative method to respond and adapt to the impacts of climate change. It focuses on how to integrate climate change risks and hazards with information and indicators that describe both the exposure and sensitivity of the highly-dense urban areas of Asian megacities such as Ho Chi Minh City (HCMC), Vietnam. As a result in part it will show how areas for mitigation of climate change risks can be determined.

2 The Impacts of Climate Change on Ho Chi Minh City

Vietnam is one of the countries which already suffer from the effects of climate change. Extreme weather events and floods have increased (Ho Phi Long, 2007). In particular, HCMC is suffering from extreme heat islands due to the urban structure. The increase of floods is due to the tidal flow superimposed with heavy rain

M.-W. Han and J. Lee (Eds.): EKC 2010, SPPHY 138, pp. 1–7.

events. Other causes are the rise in sea level and the outdated and overloaded sewage system of the city (Duong Van Truc Doan Canh & 2006). The flooding problems are from consequences of continuing urbanization, e.g. by booming construction in flood prone areas which were in the past as retention areas available (Eckert & Weibel 2009).

As land sealing results in increased overflow and extreme flooding, current status and minimization potential of land sealing should be determined as one of the important objectives of the project. Through the statistical downscaling of global climate models, the future impacts of global change on South Vietnam using IPPC high emission scenarios have been predicted as

- **Sea Level Rise:** Averaged presumed sea level rise by 2050, 18.5 cm, rising to 52.9 cm in 2100. While worst case scenarios are 33.4 cm by 2050 and 101.7 cm in 2100.
- According to many international reviews on the potential impacts of sea level rise, Vietnam ranks highly among the most affected countries (DASGUPTA ET. AL. 2007).
- **Temperature:** Predicted to be 4.0 °C to 4.5 °C higher in 2100, relative to 1990 levels.
- **Precipitation:** An overall 5-10% increase is envisaged from high emission scenarios; while during the rainy season a 10-20% increase is envisaged. During the dry season, a decrease of 20% is foreseen.

For a densely populated and highly vulnerable mega-urban region such as HCMC, timely Adaptation planning to the unavoidable effects of climate change is necessary and essential.

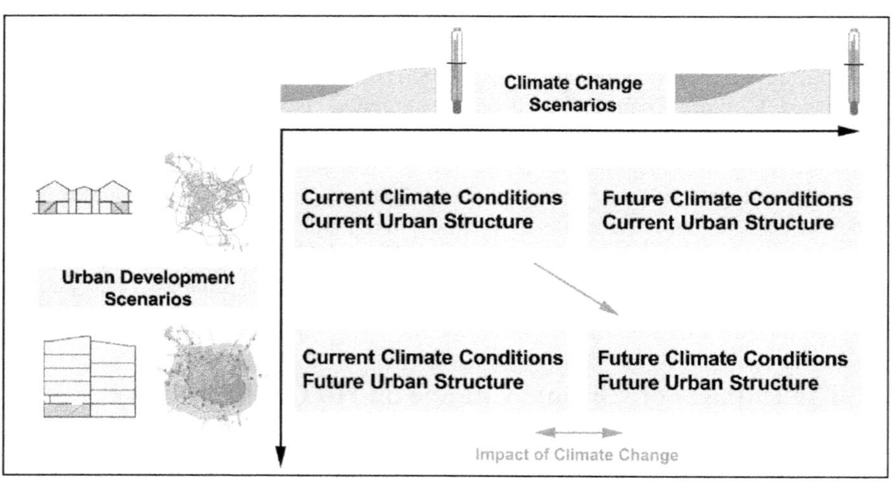

Fig. 1 Climate Change and Urban Development Scenarios

3 Downscaling Climate Change Impact on Urban Scale

Ho Chi Minh City´s (HCMC) settlements are integrated in an urban system that is affected by a number of internal and external pressures. Therefore the impacts of climate change on the city, its settlements and infrastructure should be assessed in the context of this complexity. Vulnerability to climate change will vary considerably from settlement to settlement and even within settlements. The location, urban structure, dominant building type, socio-economic characteristics and institutional capacity are key factors that affect vulnerability and adaptive capacity of a settlement in the mega-urban region. The consequences of climate change and the build-up area development of the city will be influenced by the economic, social and technological conditions, which will for HCMC be very different from those of today.(Fig.1) These conditions will have an effect on the vulnerability of HCMC´s future settlement structure to climate change impacts, by influencing the future 'adaptive capacity' – the ability of the biophysical urban structure to adapt to climate change impacts by increasing their resilience to climate change effects. The resultant UST´s are integrated within a planning information systems, to document the direct climate forced consequences at the urban scale, as well as urban feedbacks to the regional climate.

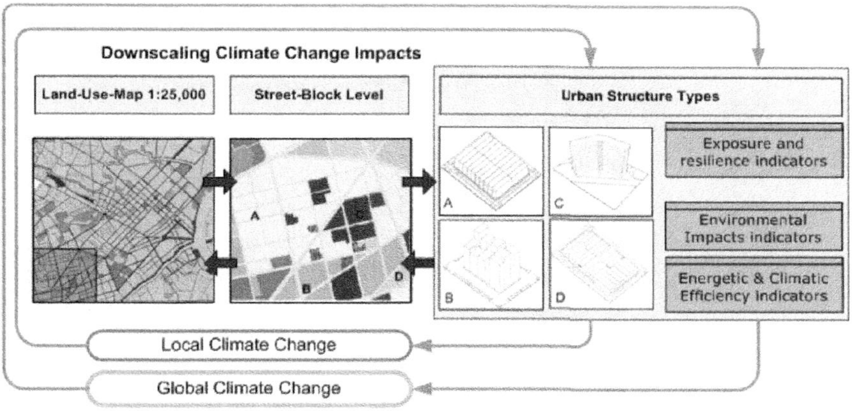

Fig. 2 Downscaling Climate Change Impacts to the Urban Scale (Source: Storch, Downes and Moon, 2009)

4 Urban Structure Types

Different discipline-specific methodological approaches to assess the vulnerability of the "urban environment" to climate change risks require a commonly accepted spatial working basis, which can ensure that the resulting heterogeneous investigations can be trans-disciplinarily integrated by using an adequate spatially explicit classification.

Typologies in spatial science are the essential way to reduce complexity of urban phenomena. An 'urban typology' provides a uniform methodological and spatial framework for the different tasks within the multi-criteria network of the research project. The Urban Structure Typology approach originated through the utilization of remotely sensed imagery to capture the urban physical form and its comprising building structures for the field of urban morphological analysis. (MOON ET. AL. 2009).

It defines primarily an urban block with physiogonomic homogeneous characteristics, which may or may not comprise varying building typologies. The block size and form is dependent upon the transportation or surface water networks that frame the block, as well as the formal or informal Layout of the building typologies and Open space. For example Shophouse based development can be clearly recognized and differed from Villa based the development or other Development patterns like High-Dense Apartments. Here the urban infrastructure exerts a dominate role. Other differentiations are based upon land uses, orientation, structure of density and sealing material. The typology approach ensures that data integration of different sources with their original specific spatial/temporal resolutions and thematic contents can be operationally integrated into the GIS environment. Different resilience and exposure indicator, environmental impact indicators and energetic / climatic efficiency indicators of Urban Structure Type (UST) provides a refined land use Pattern on the Polygons of official Land Use Map (LUM) at a scale of 1:25000. (Fig.3) The UST are identified in 4 categories (residential,

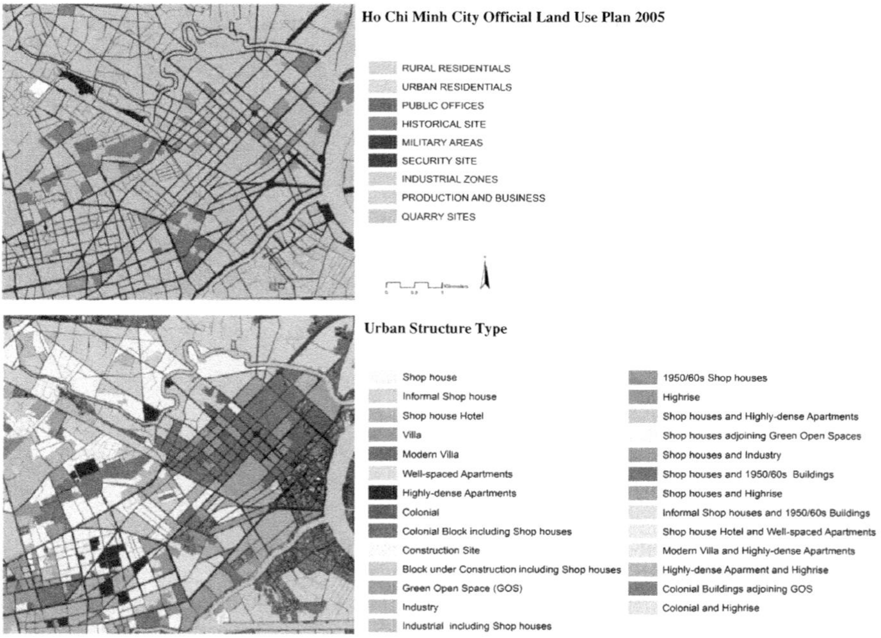

Fig. 3 Comparison of Urban Land Use and Urban Structure Types in the CBD of HCMC

industrial, public, green sector) and 83 types. The urban structure types are coded in ascending fashion for the ease of modification and adaptability of the systems through further findings or for modelling applications i.e. GIS analysis.

5 Method and the First Analysis of LUM and UST

The proposed concept of analysis for determining potential area is based on comparing of UST and LUM. According to the features of USTs, the real-using of LUM (plan) can be distinguished. Therefore results also enable to assess the status of planning, which most planning departments annoyed or neglected to show the aberration. The arrangement of UST designations against the LUM provides a matrix. By arranging planning in the horizontal direction and the UST in the vertical direction, one can get an overview on the planned and deviated areas. Hereby not only the proportion, but also the size of the areas is to be identified, in order to develop appropriate measures. The field 3 (Fig.4) shows land use in accordance with the planning. The field 2 represents informal or illegal development in green areas such as protected agricultural area, water or forest protection area.

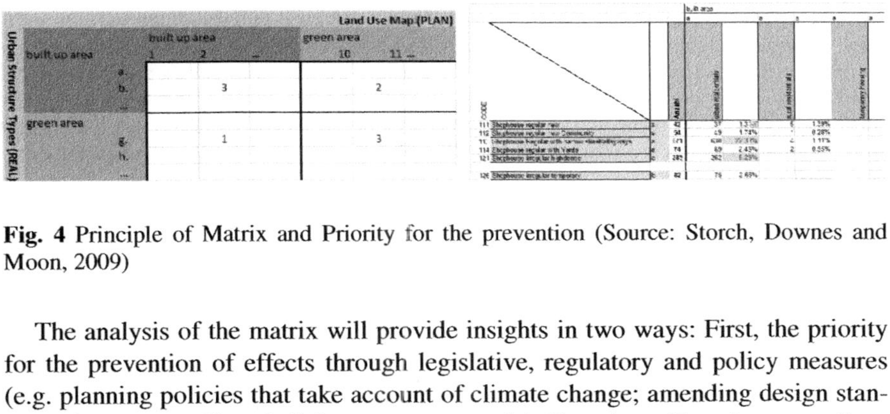

Fig. 4 Principle of Matrix and Priority for the prevention (Source: Storch, Downes and Moon, 2009)

The analysis of the matrix will provide insights in two ways: First, the priority for the prevention of effects through legislative, regulatory and policy measures (e.g. planning policies that take account of climate change; amending design standards for more resilient building structures and (re)location of housing away from high-risk areas or from ecological areas which are of importance in respect to the climatic aspect). Second, prevention of effects through combined structural and technological measures (e.g. the construction of resilient buildings and housing structures; possibilities to adapt renewable energy source; increase in water storage capacity of new urban development; possible combination of building types for avoiding heat island effect, etc.).

The results will show that the temporal comparison of the spatial balance can highlight the development trends.

Based on the matrix, results of the analysis are as following:

- Not intended (deviation from planning) land use of 1424 blocks, in an area of 21km^2
- The field 1 indicates the most important potential area for the application of optimized measures. There exist in total 355 blocks in an area of 5.7 km² comprising 5.4% of the Central Business District.
- Furthermore, not intended land use(in the Field 2) is resulted by plantation, freshwater aquaculture, remaining water paddy, natural grassland, other annual crops, other perennial crops, and other annual crops. Within such crop land, settlements are often compulsory for the agricultural workers. Along the streets, new informally managed developments are already observable.

Acknowlodgement. The research project 'Integrative Urban and Environmental Planning for Adaptation Ho Chi Minh City to Climate Change - Sustainable Strategies for Climate-Oriented Urban Structures, Energy-Efficient Housing Typologies and Comprehensive Environmental Protection for Megacities of Tomorrow' is financed as part of the new research programme 'Sustainable Development of the Megacities of Tomorrow' by the German Federal Ministry of Education and Research (BMBF). Following an evaluation of the preoperational phase of the project 'Sustainable Housing Policies for Megacities of Tomorrow - the Balance of Urban Growth and Redevelopment in Ho Chi Minh City' (2005-2008), the main phase of the BMBF-research programme, now focuses on "energy- and climate-efficient structures in urban growth centres", initiated mid 2008 and will run up to 2013 (www.emerging-megacities.org). Coordinated by the lead partner Brandenburg University of Technology Cottbus, a multi-disciplinary German, European and Vietnamese team, aims to develop an integrated "Adaptation Planning Framework for Climate Change in the Urban Environment of Ho Chi Minh City" (www.megacity-hcmc.org). Die Arbeit stellt eine Teilaufgabe der Work Packages(WP) „WP 1 - Adaptation Planning Framework.

References

Clark, G.E., Moser, S.C., Ratick, S.J., Dow, K., Meyer, W.B., Emani, S., Jin, W., Kasperson, J.X., Kasperson, R.E., Schwarz, H.E.: Assessing the vulnerability of coastal communities to extreme storms: the case of Revere, Mitigation and Adaptation Strategies for Global Change, MA., USA, vol. 3, pp. 59–82 (1998)

Cutter, S.L.: Vulnerability to environmental hazards. Progress in Human Geography 20(4), 529–539 (1996)

Das Gupta, S., Laplante, B., Meisner, C., Wheeler, D., Yan, J.: The Impact of Sea Level Rise on Developing Countries: A Comparative Analysis. In: World Bank Policy Research Working Paper 4136. World Bank, Washington, DC (2007)

Van Truc, D., Canh, D.: Gradual application of sustainable urban drainage system to Flood by overflow-rain and protect environment and resources in Ho Chi Minh City. In: The science and practise of flood disaster management in urbanizing Monsoon Asia, International Workshop, Chiang Mai, Thailand, April 4-5 (2007)

Eckert, R., Waibel, M.: Climate Change and Challenges for the Urban Development of Ho Chi Minh City- Vienam. Pacific News Nr.31, S.19 (2009)

Füssel, H.M.: Vulnerability: a generally applicable conceptual framework for climate change research. Glob. Environ. Change 17, 155–167 (2007)

Haggag, M.A., Ayad, H.M.: The urban structural units method: a basis for evaluating environmental prospects for sustainable development. Urban Design International 7(12), 97–108 (2002)

Phi, H.L.: Climate Change and Urban Flooding in Ho Chi Minh City. In: Proceedings of the 3rd International Conference on Climate Change and Water, Helsinki, Finland, Septmber 3-6, pp. 194–199 (2007)

Moon, K., Downes, N., Rujner, H., Storch, H.: Adaptation of the Urban Structure Type Approach for Vulnerability Assessment of Climate Change Risks in Ho Chi Minh City. In: Proceedings of the 45th ISOCARP International Congress 2009, Porto, Portugal (2009)

Storch, H., Downes, N., Moon, K.: Climate Change and the Resilience of Megacities in South-East-Asia Creating Risk-Based Climate Change Information for Ho Chi Minh City's Settlements. In: Proc. REAL CORP 2009, Assessment. Environmental Protection in the European Union, April 22-25, vol. 3, pp. 215–226. Springer, Heidelberg (2009)

Anthropogenic Climate Change in the Zero-Carbon Era

Chihak Ahn[1] and Nick E.B. Cowern[1,2,*]

[1] School of Electrical, Electronic and Computer Engineering, Newcastle University, Newcastle upon Tyne, NE1 7RU, UK
[2] Institute for Research on Environment and Sustainability, Newcastle University, Newcastle upon Tyne, NE1 7RU, UK
nick.cowernt@ncl.ac.uk

Abstract. Global warming is a result of 'temperature forcing', the net imbalance between energy fluxes entering and leaving the climate system and arising within it. At present humanity introduces temperature forcing through greenhouse gas emissions, agriculture, and thermal emissions from fuel burning. Up to now climate projections, based on projected GHG emissions and neglecting thermal emissions, typically foresee maximum forcing at a date occurring from midcentury onwards, followed by a slow decline due to elimination of carbon emissions. However, under reasonable scenarios of growth in primary energy use, even if we switch completely to generation by zero-carbon fuel burning (nuclear or fossil with carbon capture) temperature forcing will be sustained and even increase through the second half of the century as a result of the additional heat injected into the climate system. A potential solution to this problem is to develop energy generation technologies that remove heat from the climate system, or as a temporary solution 'dump' heat in the deep ocean. Two such technologies, both relying on solar energy, are discussed in this paper.

After several decades of significant global warming and of intensive climate research it is clear that anthropogenic CO_2 emissions must be largely eliminated during the first half of this century in order to minimise the risk of dangerous climate change [1-4]. However, what has not been widely understood is the likely climate impact of thermal emissions from power generation and use. The problem has been raised by Chaisson [5] in the context of warming over the next few hundred years, with the suggestion that it may impact significantly by mid-next century as a result of growth in human energy use, though quantitative temperature forcings were not calculated. Here we show that thermal emissions may impact on global climate much sooner, during the course of this century, as a result of projected growth in energy use and changes in the way that energy is generated and used.

Energy technologies such as nuclear (fission or fusion), fossil fuels and geothermal power plants are human-made sources of heat energy injected into Earth's

---

* Corresponding author.

M.-W. Han and J. Lee (Eds.): EKC 2010, SPPHY 138, pp. 9–19.
springerlink.com

climate system. This heat is added directly to Earth's thermal budget and contributes to global warming. In contrast, most renewable energy technologies, such as wind, wave and tidal power, harvest energy from Earth's dissipative systems and thus do not directly add to Earth's heat budget. Solar electricity generation, a promising and fast-expanding energy technology, acts in a more complex way because it exploits an existing energy flow (incoming solar radiation) but in so doing, for the purpose of efficient energy generation, typically lowers the albedo of Earth's surface at the solar collector location, thus adding to Earth's heat budget. Still, the thermal impact of solar generation may be less than that of heat-based energy sources like nuclear and geothermal power, because solar collectors take the place of terrain which was already absorbing a significant proportion, typically from 60–90%, of incident solar energy.

The flow of human-made heat into the climate system plays only a small part in present-day global warming, but as the world moves to a low-carbon energy economy increasingly dominated by electricity generation, this transition, together with expected growth in consumption, will lead to serious warming effects in addition to those previously caused by human-made CO_2. At present, the reduction of CO_2 emissions must be humanity's paramount concern, and any cost-effective zero-carbon technology (including carbon capture and storage, nuclear or geothermal energy) is preferable to a carbon emitting one. However by midcentury technologies will need to be in place to generate usable energy without significant thermal emissions integrated over the full cycle of generation, transmission and energy consumption. This is a key consideration for long-range funding choices between competing energy technologies such as fusion and renewables, which could potentially contribute substantial proportions of the world's energy supply from midcentury onwards.

In this paper we first consider the global temperature forcing arising assuming humanity's energy is obtained from heat sources such as fuel burning and geothermal power, for which primary energy is converted 100% into heat. The resultant heat-driven forcing is compared to a typical estimate of CO_2 forcing assuming responsible measures are taken to control CO_2 emissions [2]. The heat-driven forcing is shown to contribute increasingly to total forcing, suppressing the decline in forcing from midcentury onwards which climate scientists have assumed will occur after CO_2 emissions have fallen significantly [1]. We then turn to an evaluation of the likely impact of the growing technology of solar energy generation, considering various scenarios based on different types of solar technology. We find that one solar technology option could potentially generate power *and produce a net negative temperature forcing*. Finally, we consider the impact of ocean thermal energy conversion, which could help cool a portion of the climate system (the near-surface tropical ocean) which is currently driving rapid global warming and sea-level rise.

In the following paragraphs we consider the effects of growth in primary energy use. Currently primary energy use is about 15 TW, is increasing at about 2%/yr [6] and apart from short-term variations is likely to continue increasing at a similar rate for the foreseeable future. Figure 1 shows the CO_2 forcing (black solid curve) resulting from emissions in the 'Coal Phase-Out' scenario of Kharecha and

Hansen [2], and the sum of CO_2 forcing and thermal forcing due to primary energy use with 2% growth (red dashed curve). Instead of peaking and subsequently decreasing from midcentury onwards as in Ref. [2], the total forcing from CO_2 and thermal emissions stabilises for about 80 years at ~2.6 W/m^2, corresponding to an equilibrium temperature rise of nearly 4°C [3] if sustained, and then the forcing rapidly rises further. Based on virtually all accepted climate models, this would lead Earth into a period of dangerous climate change.

Also shown in figure 1 is an alternative scenario which follows Chaisson's suggestion [5] to attempt a more realistic distribution of growth rates between OECD and non-OECD countries. World primary energy use is currently divided almost equally between OECD and non-OECD countries, while non-OECD population growth and consumption per head are increasing faster than in the OECD. We assume that OECD energy use will grow at 1%/yr while non-OECD energy use will grow at 5%/yr until per-capita consumption equals that in the OECD, after which all growth will continue at 1%/yr [5]. To simplify analysis we treat OECD and non-OECD countries as two separate blocks, so that transfer of individual countries from one block to the other is neglected. In this scenario the non-OECD 'block' catches up by midcentury and subsequent growth is at 1%/yr. The result (orange chained curve) is an increase in total forcing by about 5% by year 2040 relative to CO_2 forcing alone, and for the following 150 years an almost flat plateau in total forcing, suppressing the gradual decrease in forcing predicted for the case of CO_2 emissions alone. Both the 2% growth scenario and the '5%+1%' growth scenario lead to accelerated global temperature rise compared to current climate simulations, and in both cases the result is drastically worse than recent predictions of long-term global warming based in residual long-term levels of atmospheric CO_2 [7].

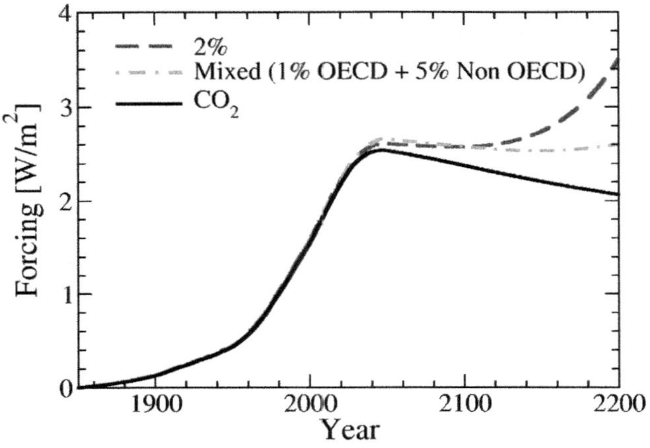

Fig. 1 Global temperature forcing resulting from CO_2 emissions in the 'Coal Phase-Out' scenario of Kharecha and Hansen. Black solid curve - result without taking thermal emissions into account. Red dashed curve - result assuming 2% growth in primary energy use. Orange chained curve – result using the '1%+5%' scenario discussed in the text.

Future temperature forcing may be larger than the projected values in figure 1, since we have not yet taken into account the likely effects of a transition to electricity-based energy consumption, which will be needed if we are to utilise zero-carbon fuel-based energy sources. Such a transition will increase the consumption of primary energy needed to supply a given amount of useful energy to consumers, as efficiency in the delivery of energy falls to values in the region of 35% [1] associated with electricity generation and supply. As we shall show later in the paper, this leads to a substantial further increase in global temperature forcing in the case of fuel-based and geothermal energy sources.

If heat-based energy sources could be fully supplanted by 'passive' renewables such as wind, wave or tidal energy, thermal emissions would be much less significant – only heat generated in plant construction and maintenance, and possibly second-order changes in the climate system owing to perturbations of natural energy flows by these energy conversion systems, would play a role. However, current forecasts suggest that such energy sources, while important, cannot supply all of humanity's energy needs, and much research, technology development and manufacturing is currently being devoted to solar-based electricity generation, with the photovoltaic (PV) market growing at a remarkable projected rate of about 50%/yr according to some estimates [9].

In a recently published Solar Grand Plan, Zweibel *et al.* have proposed a strategy to transform the US energy economy from its current fossil-fuel rich mix to one dominated by solar power [10]. By 2100 the transition to a solar energy economy with modest contributions from wind and geothermal power is essentially complete as solar electrical output plus other renewables provide over 90% of the total energy supply including transport.

Fig. 2 shows schematically the impact of a solar PV-based power plant on local and global energy balance, taking into account the transmission of electrical energy away from the collector area towards distant consumers. Installation of the PV collector array increases local solar absorption because the terrain albedo, a_t, typically 0.25 – 0.45 for a hot, flat desert environment [11,12], is reduced to the effective albedo of the PV array, a_c. In most current PV technologies, $a_c \approx 0$, since effective reflectivities of solar collector surface films are currently in the range of a few per cent and falling as further technical advances are made. The net increase in solar flux absorbed within the collector area, A , is then $\delta\phi = (a_t - a_c)\phi$, where ϕ is the incident solar flux. The amount of this energy flux transmitted as electricity to load, $\varepsilon\phi$, is dependent on the solar conversion efficiency of the installed collector plant, ε , however at remote locations virtually all of this transmitted energy ultimately converts to heat, thus the net increase in global heat input per second, ΔW_G , is simply

[1] The maximum possible efficiency for future nuclear technology appears to be in the region of 50%, see reference [8]. Here we have used the more conventionally accepted efficiency value of ~ 35% for nuclear or fossil fuel generation.

$$\Delta W_G = (a_t - a_c)\phi A \qquad (1)$$

where A is the global area covered by solar collectors. We now address the issue of global temperature forcing represented by this equation.

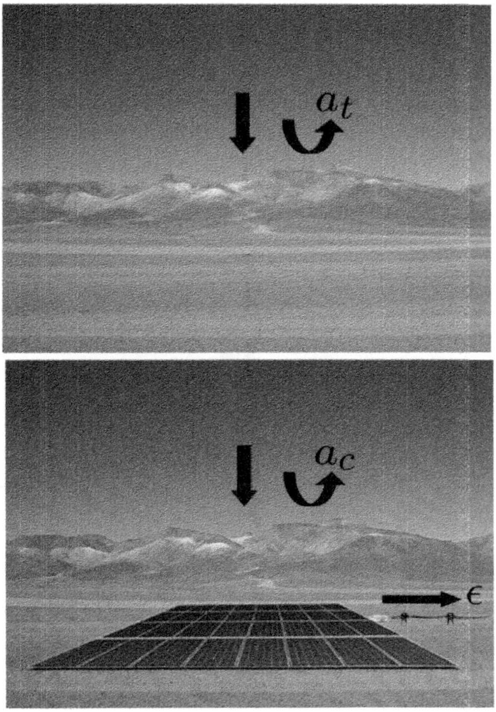

Fig. 2 Schematic view of a solar PV-based power plant before and after installation. The albedo of the rectangular area changes from a_t, that of the local terrain, to a_c, the albedo of the PV array, leading to an change in the energy flux entering this area. Local climate is further modified by the outflow of electrical energy to the transmission network or storage site, however this relocation of energy does not alter net global temperature forcing since virtually all the relocated energy ultimately transforms into heat.

In order to do this we apply the '1%+5%/yr' growth scenario to the output of energy supplied to users. Following Zweibel *et al.*, the solar contribution is assumed to grow by about 6%/yr between 2020 and 2100, after which it becomes the dominant energy source and grows by 1%/yr constrained by our assumption on total energy use. Excess heat production is calculated assuming an average original terrain albedo of 0.3, a collector albedo of 0.0 (100% absorption), and a solar efficiency of 20%. The resulting global temperature forcing is plotted as the orange chained curve in Fig. 3. For comparison the red dashed curve shows the temperature forcing if the same growth in electrical generation is provided by burning fuels (nuclear or fossil with carbon capture and storage) and/or geothermal energy. Qualitatively the forcing is similar to that caused by solar energy, but the effect is larger since all of the primary energy used for electricity production contributes to global warming. The advantage in the solar case arises from the factor $a_t - a_c$ in equation (1); the excess heat produced by solar power is equal to the radiant energy absorbed by the collector, *minus* that which the original terrain would have absorbed as part of Earth's normal radiation balance.

One way to compensate for the temperature forcing caused by solar power is to use a form of geoengineering [13] known as 'albedo engineering' in which an area of relatively low albedo is replaced by a high-albedo surface. The area of high-albedo surface needed to compensate this is smaller than the solar collection area, thus if solar collectors are economically feasible, this additional technology is presumably feasible as well.

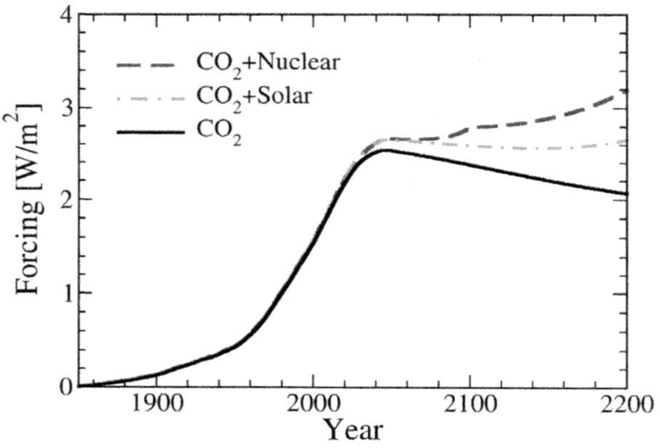

Fig. 3 Global temperature forcing in the '1%+5%' scenario for growth in energy use, assuming a transition to all energy use in the form of electricity by 2100. The orange chained curve corresponds to electricity generated from solar, assuming 20% efficiency and a 30% decrease in surface albedo on installation of a solar collector array. The red dashed curve corresponds to electricity generated by zero-carbon fuel-burning, where all heat released from burning is a contribution to temperature forcing.

A more direct approach incorporates albedo engineering directly into PV technology by backing a thin active PV layer with a reflective substrate. Given a suitable choice of PV material and device structure, this enables much of the unused energy in the long-wavelength spectrum of sunlight (below the semiconducting band gap) to be reflected back out of the entry surface of the PV cell, thus raising its effective albedo. While a full evaluation of this effect is beyond the scope of this paper, we make some simplified model calculations. We consider an idealised PV cell with a simple active layer 1 (coloured blue) and a substrate layer 2 (coloured red) as illustrated in Fig. 4a. Photons with energy E above the bandgap E_g of layer 1 are assumed to be converted with 100% quantum efficiency, i.e. with energy efficiency E_g / E, while a proportion A of photons below the bandgap energy are assumed to be absorbed in layer 2, i.e. not reflected back through the entry surface or transmitted through layer 2. For simplicity we assume that A is independent of wavelength. The quantities R and T shown in Fig. 4a are the fractions of the sub-bandgap light reflected and transmitted, respectively, by layer 2.

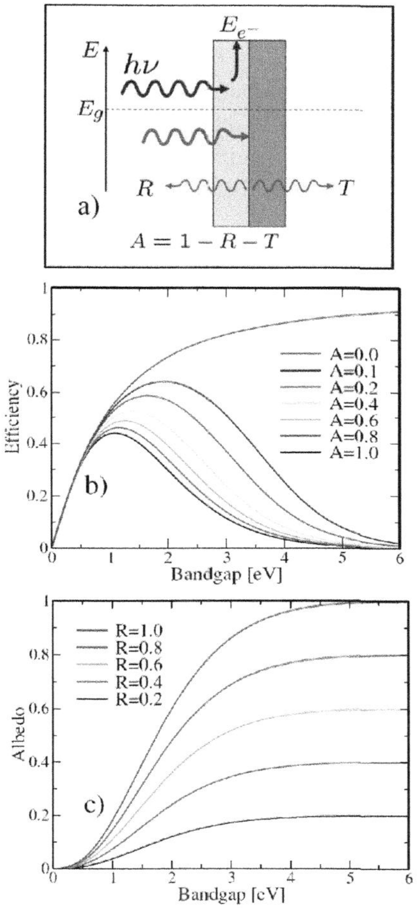

Fig. 4 Radiation fluxes and photoelectron generation in a simplified one-dimensional PV cell model. a) - model cell schematic. b) - thermal efficiency in the model PV cell. c) - approximate effective albedo of the model cell. The combination of high thermal efficiency and high albedo at bandgaps in the range from 1 eV upwards illustrates the potential of thin-film reflective PVs for large-scale solar 'cool power' generation.

In this simple model the energy absorbed by the cell is a sum of the energy converted to charge carriers in layer 1 and the energy absorbed in layer 2, i.e. $N_> \overline{E}_> + AN_< \overline{E}_<$, where N indicates the photon flux and $\overline{E}$ the mean photon energy in the relevant energy range, above (>) or below (<) the bandgap.

The 'thermal efficiency' of our model cell, defined as the fraction of this absorbed energy converted to electrical energy, is

$$\varepsilon = \frac{E_g}{\overline{E}_> + (N_< / N_>) A \overline{E}_<} \tag{2}$$

In general this efficiency is higher than that defined in terms of electrical output divided by incident solar energy flux, because the denominator includes only the absorbed light. For a reflective solar cell, i.e. one with no transmission through the back of the cell, *A=1-R* and the effective albedo of the cell is given by the ratio of reflected to incident energy fluxes,

$$a_c = \frac{RN_< \overline{E}_<}{N_> \overline{E}_> + N_< \overline{E}_<} \tag{3}$$

Fig. 4b shows the thermal efficiencies of cells with values of *A* in the range from 0 to 1, and Fig. 4c shows the corresponding effective albedo values. For bandgap values ranging from that of crystalline Si (1.1 eV) up to 2.5 eV, and for A=0.1, thermal efficiency is in the range ~55-65%, higher than can be achieved with conventional energy technologies. In addition, cell albedo may well be higher than terrain albedo, offering a theoretical possibility that PV technology could produce a negative temperature forcing supporting global cooling.

As an example, we present the time evolution of temperature forcing for a global solar grand plan using reflective PV technology with R=0.9 (A=0.1). The impact of such a technology is shown in Fig. 5, the higher value (orange chained curve) corresponding to the silicon bandgap and the lower value (green dotted curve) corresponding to a wide-bandgap PV material with $E_g = 2$ eV. The curves for net global temperature forcing using 1.1 eV or 2 eV bandgap now lie close to the prediction for CO_2 forcing, and for the higher bandgap, total forcing is actually *below* the CO_2 forcing curve because the PV albedo of 0.54 exceeds terrain albedo, thus producing a net cooling effect. Although the trend of these results is clear the curves for the different bandgaps in Fig. 5 are not quantitatively exact given the simplifications used in our analysis. Further work to refine these predictions will be reported in a later paper.

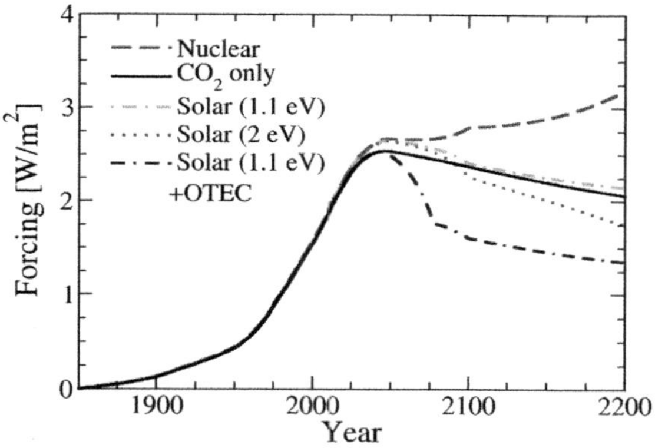

Fig. 5 Global temperature forcing versus time as a result of CO_2 emissions (black solid curve), CO_2 emissions plus thermal effects from solar generation using reflective thin-film PVs with bandgaps of 1.1 eV (orange chained curve) and 2 eV (green dotted curve). For the 1.1 eV bandgap value, collector albedo is 0.21, slightly below our assumed terrain albedo of 0.3, leading to a small positive forcing contribution. However, for the 2 eV bandgap, collector albedo is 0.54 thus introducing a negative temperature forcing contribution once PV becomes the dominant contribution to energy generation. Finally, we show the impact of ocean thermal energy conversion, assuming the same growth rate as solar up to a maximum of 20 TW (blue chained curve). In this case, temperature forcing is approximately halved over the 100 years following peak forcing, substantially reducing the magnitude of the forcing 'impulse' leading to long-term global warming.

Finally, we show the potential impact of the developing technology of ocean thermal energy conversion (OTEC), which generates electricity by pumping heat from warm ocean-surface waters to the cooler, deeper ocean [14]. Here the heat sunk per output electrical energy, $1/\varepsilon$, is high due to the relatively low thermodynamic conversion efficiency, ε, of heat pump technology. This determines the amount of heat taken from the climate system after accounting for use of electrical output power,

$$-\Delta W_G = P(\varepsilon^{-1} - 1) \tag{4}$$

where P is global electrical output and the term -1 accounts for heat created by electricity use.

We assume as in most climate models that the ocean surface couples strongly (i.e. rapidly) to the atmosphere, we treat the slow transport of buried heat back to the ocean surface as negligible on the time scale of our predictions, and we assume $\varepsilon = 0.05$, close to the maximum expected from OTEC model simulations [15], so providing a conservative estimate of ε^{-1}. OTEC generation is assumed to grow at the same rate as solar, up to a maximum of 20 TW which is probably an ultimate limit for energy generation with this technology.

The result, shown in Fig. 5 by the blue chained curve, is dramatic. Even a substantially smaller contribution of OTEC to global energy generation, producing a proportionately smaller negative temperature forcing, could be an important contribution to stabilising global surface temperature. The key principle here is that heat is pumped to the deep ocean much faster than is achieved by natural ocean heat transport processes. In this way, OTEC, with an appropriate magnitude and spatial distribution of generating capacity, could help control the rising trend of ocean surface temperature and temperature gradients which are driving fast, potentially dangerous, climate feedbacks.

While dumping thermal energy into the deep ocean is not a long-term solution, it potentially helps limit the 'impulse' that anthropogenic forcing over the next century will introduce towards longer-term global warming. Reflective thin film solar technology may offer a better solution in the longer term by actually expelling energy from the climate system.

We have shown that thermal effects from human energy consumption will play an increasingly significant role in global temperature forcing in the medium-term future. Consequently it is important to discriminate between energy sources that inject heat into Earth's climate system (fuel burning and geothermal energy), those that rely passively on Earth's dissipative systems (wind, wave, tidal energy), and those that may potentially remove heat energy (suitably chosen solar technology, OTEC, and perhaps other future technologies). Correct technology choices will reduce the magnitude and time period of future global warming caused by current CO_2 emissions. Conversely, nuclear fusion, which may potentially come on stream as a significant energy source several decades hence, will be too late as a replacement for CO_2-emitting technologies, and will not meet contemporaneous thermal emissions criteria for a sustainable global environment unless its thermal effects are compensated by additional geoengineering schemes. We suggest a re-evaluation by the science and engineering communities, taking thermal cycle analysis into account, so that the most promising future technologies for zero-carbon, thermally-compliant energy generation can be targeted for research and development during the next decade.

Acknowledgement. We thank N.S. Bennett and S.J. Bull for discussions.

References

[1] IPCC Fourth Assessment Report 2008 Climate Change Synthesis Report (2007), http://www.ipcc.ch/pdf/assessment-report/ar4/syr/ar4_syr.pdf
[2] Kharecha, P.A., Hansen, J.E.: Global Biogeochem. Cycles 22, GB3012 (2008)
[3] Hansen, J., et al.: Open Atmos. Sci. J. 2, 217 (2008)
[4] Hansen, J., et al.: Target atmospheric CO2 : Supplementary Material, preprint arxiv:0804.1135 (2008)
[5] Chaisson, E.J.: Eos, Transactions. American Geophysical Union 89, 253 (2008)
[6] BP Statistical Review of World Energy (June 2008)

[7] Inman, M.: Nature Reports Climate Change 2, 156 (2008)
[8] Wong, A.: Trans. Am. Nucl. Soc. 68, 380 (1993)
[9] Bank Sarasin Sustainability Research Team 2008 Solar energy (2008) (Bank Sarasin Report)
[10] Zweibel, K., Mason, J., Fthenakis, V.: Sci. Am. 298, 48 (2008)
[11] Claussen, M.: Climate Dynamics 13, 247 (1997)
[12] Knorr, W., Schnitzler, K.-G., Govaerts, Y.: Geophys. Res. Lett. 28, 3489 (2001)
[13] Keith, D.W.: Ann. Rev. Energy and Env. 25, 245 (2000)
[14] Vega, L.A.: Marine Technol. Soc. J. 36, 25 (2002)
[15] Berger, L.R., Berger, J.A.: Appl. Environ. Microbiol. 51, 1186 (1996)

Design of LCL Filter for Renewable Energy Sources Using Bacterial Foraging Optimization

Jae Hon Cho[1] and Dong Hwa Kim[2,*]

[1] Chungbuk National University, 410 Seongbong-Ro Heungduk-Gu, Cheongju Chungbuk, Korea
[2] Hanbat National University, San 16-1, DuckMyoung-Dong, Yuseong-Gu, Daejeon, Korea
kimdh@hanbat.ac.kr

Abstract. As the traditional resources have become rare, renewable energy sources are developing quickly. The grid connected renewable sources is one of the most importance problem in smart grid fields. To reduce harmonic in this grid, many filter techniques have been developed. Compared traditional L filter, a LCL filter is more effective on reducing harmonic distortion at switch frequency. So, it is important to choose the LCL filter parameters to achieve good filtering effect. In this paper, a design method of LCL filter by bacterial foraging optimization is proposed. Simulation result and calculate data are provided to prove that the proposed method is more effective and simple than traditional methods.

1 Introduction

Voltage-source PWM method has many advantages such as bidirectional power flow, controllable power factor and sinusoidal input current. A PWM converter with higher switching frequency will result in smaller LC filter size. However, switching frequency is generally limited in high power applications. As an alternative solution, LCL filter is more attractive for two reasons: [1]

It has better attenuation than LC filter given the similar size.

LCL filter provides inductive output at the grid interconnection point to prevent inrush current compared to LC filter.

In 2005, Marco Liserre presented a step by step design method to design the LCL filter [2]. But the method is complicated.

Recently, artificial intelligence has become a popular search technique used in computing to find exact or approximate solutions to optimization and search problems. There are many kinds of artificial intelligence and Genetic Algorithm (GA) is used mostly[3]. Genetic algorithms are a particular class of evolutionary algorithms the use techniques inspired by evolutionary biology such as inheritance, mutation, selection, and crossover [4]. It becomes successful because of its concise arithmetic describing. On the other hand, as natural selection tends to eliminate animals with poor foraging strategies through methods for locating, handling, and ingesting food and favor the propagation of genes of those animals that have

* Corresponding author.

M.-W. Han and J. Lee (Eds.): EKC 2010, SPPHY 138, pp. 21–28.
springerlink.com

successful foraging strategies. Based on this conception, Passino proposed an optimization technique known as the Bacterial Foraging Optimization (BFO). The BFO mimicking biological bacterial food-searching behavior has been applied in the field of optimization. Optimization methods inspired by bacteria can be categorized based on the chemotaxis algorithm and bacterial foraging algorithm. The chemotaxis algorithm was proposed for analogy to the way bacteria to react chemoattractants in concentration gradients. The bacterial foraging algorithm is based on bacterial chemotaxis, reproduction and elimination-dispersal events and has been variously applied to some problems such as optimal control design [5], harmonic estimation [6], transmission loss reduction [7], active power filter synthesis[8], and learning of artificial neural networks [9]. This paper presents a new design method of LCL filter using bacterial foraging optimization.

2 Simplified Design Principle of LCL Filter

The Three-phase Voltage Source PWM converter with LCL filter is shown in Fig.1. The filter has three unknowns, L1,Cf and L2. In the following, three considerations that lead to three equations for determining the three unknowns will be discussed.

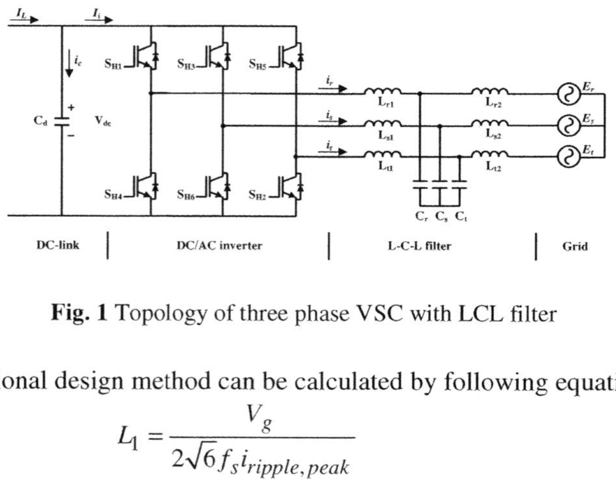

Fig. 1 Topology of three phase VSC with LCL filter

The traditional design method can be calculated by following equations[4].

$$L_1 = \frac{V_g}{2\sqrt{6} f_s i_{ripple,peak}} \quad (1)$$

$$C_f \le 0.05 C_b \quad (2)$$

$$C_b = \frac{1}{(\omega_n Z_b)} \quad (3)$$

$$Z_b = \frac{V_{gLL}^2}{P_n} \quad (4)$$

V_g is the RMS value of grid voltage, f_s is inverter switching frequency, is the base capacitance, Z_b is base impedance, V_{gLL} is grid line voltage and P_n is inverter rated power.

$$C_f = 0.025C_b \tag{5}$$

The grid side inductance of LCL filter L_2 is computed as follow.

$$L_2 = 0.8L_1 \tag{6}$$

3 Bacterial Foraging Optimization

Search and optimal foraging of animals can be used for solving engineering problems. To perform a social foraging, an animal needs communication capabilities and it gains advantages to exploit essentially the sensing capabilities of the group, so that the group can gang-up on larger prey, individuals can obtain protection from predators while in a group [10].

Optimization Function for the Bacterial foraging

For applying the bacterial foraging to optimization problem, conventional BFO was described as following [11].

In the minimal problem, the main goal of the BFO based algorithm is to find the minimum of $J(\theta), \theta \in R^p$ not the gradient $\nabla J(\theta)$. Here, θ is the position of a bacterium, and $J(\theta)$ denotes an attractant-repellant profile. That is, where nutrients and noxious substances are located, $J(\theta) < 0$, $J(\theta) = 0$, $J(\theta) > 0$ represent the presence of nutrients, neutral medium, and noxious substances, respectively. On the other hand, the population of bacteria can be defined by

$$P(j,k,l) = \{\theta^i(j,k,l) \mid i = 1,2,\cdots,S\} \tag{5}$$

where $\theta^i(i,j,k)$ represents the position of each member in the population of the S bacteria at the jth chemotactic step, kth reproduction step, and lth elimination-dispersal event. Let $J(i,j,k,l)$ denote the cost at the location of the ith bacterium $\theta^i(i,j,k) \in R^p$ and the bacterial position after the next chemotactic step can be represented by

$$\theta^i(j+1,k,l) = \theta^i(j,k,l) + C(i)\phi(j) \tag{7}$$

where $C(i) > 0$ is the size of the step taken in the random direction specified by the tumble. If the cost $J(i, j+1, k, l)$ at $\theta^i(j+1, k, l)$ is better than at $\theta^i(j, k, l)$, then another chemotactic step of size $C(i)$ in this same direction will be taken and repeated up to a maximum number of steps N_s which is the length of the lifetime of the bacteria as long as it continues to reduce the cost.

During the process of chemotactic, the bacterium which has searched the optimal position tries to provide an attractant or repellent signal for the swarm behaviors of a group. Function $J^i_{cc}(\phi)$, $i = 1, 2, \ldots, N$, to model the cell-to-cell attractant and a repellant effect is represented by

$$J_{cc}(\theta, P(j,k,l)) = \sum_{i=1}^{S} J^i_{cc}(\theta, \theta^i(j,k,l)) = \sum_{i=1}^{S}\left[-d_{attract}\exp\left(-\omega_{attract}\sum_{m=1}^{p}(\theta_m - \theta^i_m)^2\right)\right] + \sum_{i=1}^{S}\left[-h_{repellant}\exp\left(-\omega_{repellant}\sum_{i=1}^{p}(\theta_m - \theta^i_m)^2\right)\right] \quad (8)$$

where $\theta = [\theta_1, \ldots, \theta_p]^T$ is a point on the optimization domain, θ^i_m is the mth component of the ith bacterium position, $d_{attract}$ is the depth of the attractant released by the cell, $\omega_{attract}$ is a measure of the width of the attractant signal, $h_{repellant} = d_{attract}$ is the height of the repellant effect magnitude, and $\omega_{repellant}$ is a measure of the width of the repellant. Therefore, the final cost at the location of the ith bacterium $\theta^i(i, j, k) \in R^p$ reflecting the effect of an attractant and repellant can be defined by

$$J(i, k, k, l) + J_{cc}(\theta, P) \quad (9)$$

After chemotactic steps, a reproduction step is taken. The bacteria with the highest J values (low fitness cost) die and the other bacteria having lower values J (high fitness cost) split into two bacteria, which are placed at the same location, and then elimination-dispersal events is carried out. In these events, each bacterium in the population is subjected to elimination-dispersal with probability. Here, it is noted that each bacterium has the same predefined probability of elimination-dispersal events.

Design of LCL filter parameters by Bacterial foraging optimization

In this paper, the structure of bacterial position is presented as follows.

$$X = [L_1 \; L_2 \; C_f] \quad (10)$$

This paper uses harmonic attenuation rate which is defined is equation (11) as the fitness function. To achieve good filter effort, a low harmonic attenuation rate is required

$$\frac{i_g(h_{sw})}{i(h_{sw})} = \frac{z_{LC}^2}{\left|\omega_{res}^2 - \omega_{sw}^2\right|} \tag{11}$$

$$\omega_{res}^2 = \frac{L_T z_{LC}^2}{L_1}, \quad L_T = L_1 + L_2, \quad z_{LC} = \frac{1}{L_2 C_f}$$

$$\omega_{sw}^2 = (2\pi f_{sw})$$

4 Simulation Results

This paper uses MATLAB/SIMULINK to simulate photovoltaic(PV) generation systems. Table1 and Table 2 show simulation parameters for PV and Bacterial foraging optimization. Table 3 shows simulation results by the traditional method and BFO. Fig. 2 shows converter current and voltage by traditional method and Fig. 3 shows converter current and voltage by BFO. Table 3 and Fig. 3 show that the proposed method is better than traditional method.

Table 1 Simulation parameters

Parameter	Values
Inverter Power (Pn)	100kW
Utility line (VgLL)	600V
Utility frequency(fn)	60Hz
Switching frequency (fsw)	4.5kHz
DC rated voltage (Vdc)	1225V
Current(i)	96A (rms)

Table 2 Simulation parameters for BFO

	Parameter	Value
S	The initial bacteria population size for BPO	300
Nc	Chemotactic steps	100
Nre	The number of reproduction steps	2
Ned	The number of elimination-dispersal events	2
Ped	Elimination-dispersal with probability	1
C	The step size for a swimming bacteria	0.2

Table 3 Simulation results

Methods	L1	L2	Cf	fitness(F)
Traditional method	1.1mH	0.3mH	13.24uF	0.085
BFO	2.32mH	1.03mH	14.24uF	0.035

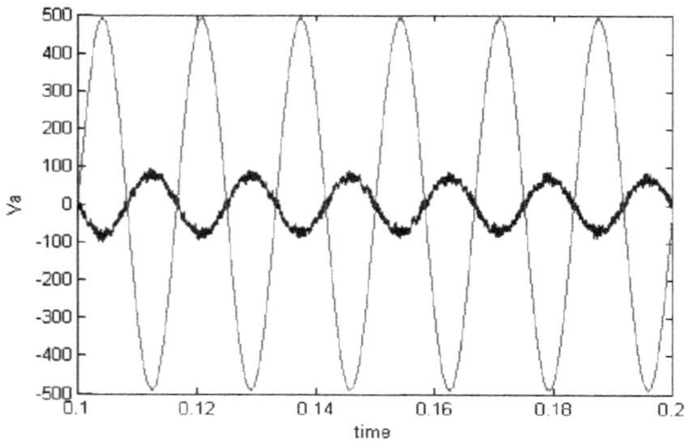

Fig. 2 Converter current and voltage by traditional method

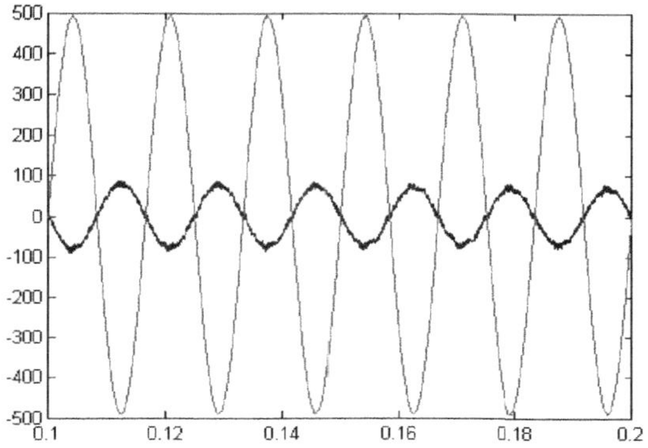

Fig. 3 Converter current and voltage by Bacterial Foraging Optimization

5 Conclusion

As the traditional resources have become rare, renewable energy sources are developing quickly. The grid connected renewable sources is one of the most importance problem in smart grid fields. To reduce harmonic in this grid, many filter techniques have been developed. In this paper, a design method of LCL filter by bacterial foraging optimization is proposed. Simulation result and calculate data are provided to prove that the proposed method is more effective and simple than traditional methods.

Acknowledgements. The work developed in this paper has been supported by the DAEDEOK INNOPLIS ("R&D Hub Cluster project").

References

[1] Wang, T.C.Y.Z., Ye, G., Sinha, X.: Output filter design for a grid-interconnected three-phase inverter. In: Power Electronics Specialist Conference, pp. 779–784 (2003)

[2] Liserre, M., Blaabjerg, F., Hansen, S.: Design and Control of an LCL-filter based Three-phase Active Rectifier. IEEE Trans. Industry Applications, 1281–1291 (2005)

[3] Wang, Z., Cui, D.: Optimization Design Based on Genetic Algorithm for Rectifier. Trans. of China Electrotechnical Society 19(5), 7–9 (2004)

[4] Sun, W., Chen, Z., Wu, X.: Intellignet Optimize Design of LCL filter for Three-Phase Voltage-Source PWM Rectifier. In: IPEMC 2009, pp. 970–974 (2009)

[5] Kim, D., Abraham, A., Cho, J.: A hybrid genetic algorithm and bacterial foraging approach for global optimization. Information Sciences 177, 3918–3937 (2007)

[6] Mishra, S.: A hybrid least square-fuzzy bacterial foraging strategy for harmonic estimation. IEEE Transactions on Evolutionary Computation (9), 61–73 (2005)

[7] Tripathy, M., Mishra, S., Lai, L.L., Zhang, Q.: Transmission loss reduction based on FACTS and bacteria foraging algorithm. In: Runarsson, T.P., Beyer, H.-G., Burke, E.K., Merelo-Guervós, J.J., Whitley, L.D., Yao, X. (eds.) PPSN 2006. LNCS, vol. 4193, p. 222. Springer, Heidelberg (2006)

[8] Mishra, S., Bhende, C.: Bacterial foraging technique-based optimized active power filter for load compensation. IEEE Transactions on Power Delivery (22), 457–465 (2007)

[9] Cho, J.H., Chun, M.G., Lee, D.J.: Parameter Optimization of Extreme Learning Machine Using Bacterial Foraging Algorithm. In: International Symposium on Advanced Intelligent Systems, pp. 742–747 (2007)

[10] O'brien, W., Browman, H., Evans, B.: Search strategies of foraging animals. American Scientist 78, 152–160 (1990)

[11] Passino, K.: Biomimicry of bacterial foraging for distributed optimization andcontrol. IEEE Control Systems Magazine (22), 52–67

A Study on TiO_2 Nanoparticle-Supported Mn_2O_3 Catalyst for Indoor Air Pollutants - Ozone and VOC Decomposition Reaction

Jongsoo Jurng[1,*], Sungmin Chin[1], and Eunseuk Park[1,2]

[1] Director of Environment Division,
Korea Institute of Science and Technology (KIST),
Seoul, Korea
jongsoo@kist.kr
[2] Graduate School, University of Seoul, Seoul, Korea

Abstract. The catalytic conversion of toluene on MnO_x/TiO_2 catalysts was examined in the presence of ozone. The MnO_x/TiO_2 catalysts were prepared by a wet-impregnation method. A set amount of manganese acetate was dissolved in distilled water. The TiO_2 nanoparticles, as the support materials in this study, were synthesized by a thermal decomposition process. The generation of TiO_2 nanoparticles by the thermal decomposition of titanium tetraisopropoxide (TTIP) was carried out in an electric tune furnace. The TiO_2 nanoparticles were characterized by X-ray diffraction, Brunauer-Emmett-Teller measurements and transmission electron microscopy. Compared to the commercial photocatalyst (P25), the TiO_2 nanoparticles prepared at 900°C had a small particle size (10.5 nm), pure anatase phase and a high degree of crystallinity with a dense polyhedral structure. The concentration of toluene was varied in the range of 20-100 ppm and the ozone to toluene ratio was increased from 1.0 to 15.0. The reaction temperature was controlled between 25-100°C and the space velocity was fixed at 20000 h^{-1}. The CO_x selectivity was enhanced considerably by the catalyst supported on TiO_2 (thermal decomposition process) after injecting ozone. The activity of the catalyst depended mainly on the support used. The catalyst synthesized by the thermal decomposition process was found to be the most active support.

1 Introduction

Volatile organic compounds (VOCs) are well-known indoor pollutants with toluene being a typical major VOC.

Manganese oxides have higher activity for the complete oxidation of VOCs and are frequently used a catalyst constituent. In addition, these are effective in the low temperature deep oxidation of toluene to carbon dioxide and water [1] and can transform ozone to active oxygen species [2,3,4]. Guo et al. [5] reported that a combination of dielectric barrier discharge and a manganese oxide/alumina/nickel foam catalyst could improve the toluene removal efficiency and carbon dioxide

* Corresponding author.

M.-W. Han and J. Lee (Eds.): EKC 2010, SPPHY 138, pp. 29–34.
springerlink.com

selectivity, and suppress byproduct formation. However, many studies on catalytic oxidation with ozone over manganese oxides focused on each support material [6,7]. In particular, the characteristics of the support material can critically affect the catalytic oxidation ability of the catalyst. Chin et al. [8] suggested that the characteristics (specific surface area and crystallinity) of the support material (TiO_2) play an important role in the catalytic conversion of 1,2dichlorobenzen using a V_2O_5/TiO_2 catalyst.

In this study, the catalytic decomposition of gaseous toluene on MnO_x/TiO_2 catalysts was carried out in the presence of ozone and the effects of the support material (TiO_2) characteristics were investigated. The microstructural characteristics of TiO_2 nanoparticles by gas phase synthesis using thermal decomposed precursor were examined as a function of the synthesis temperature and precursor bubbling temperature.

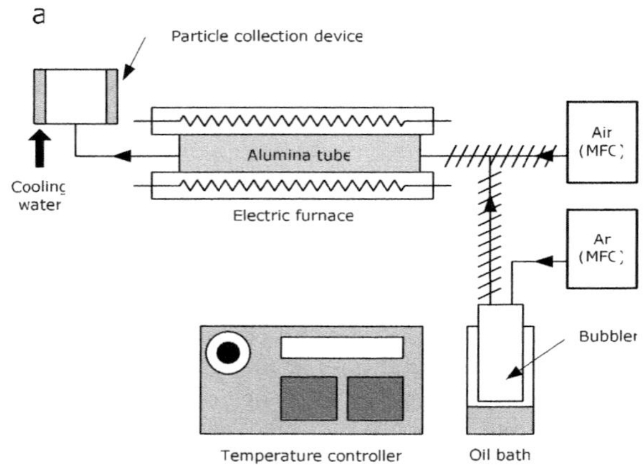

Fig. 1 Schematic diagram of the catalyst synthesis

2 Experimental Procedures

(1) Catalyst preparation

TiO_2 catalysts were prepared using a high temperature thermal decomposition of Ti precursor, as shown schematically in Fig. 1. TTIP (Titanium tetraisopropoxide, $[(CH_3)_2CHO]_4Ti$, Aldrich, >97%) was introduced into the system by bubbling argon gas (prepurified, 99.99%) through a bubbler, at a temperature controlled by a power controller. The tubing before (connected to the argon) and after (connected to the furnace system entrance) the bubbler was wrapped with heating tape to prevent condensation of the precursor. An additional inlet was connected to the reactor to supply air. An alumina tube (0.8 m in length and 0.01 m in diameter) was placed in an electric furnace controlled by a temperature controller. Downstream from the alumina tube, a water-cooled particle collection device was installed to collect the synthesized catalysts. The TTIP heating and synthesis temperatures were fixed at 95°C and 900°C, respectively. The powder samples are labeled as "P25-TiO_2" for the commercial catalyst and "TD-TiO_2" for the catalyst prepared by the thermal decomposition process.

TiO_2-supported manganese oxides (MnO_x/TiO_2) were prepared by impregnating the catalyst support with an aqueous solution containing the appropriate amount of manganese acetate ($Mn(CH_3COO)_2 \cdot 4H_2O$, JUNSEI Chemical). One hundred milliliters of distilled water was added to a 200 mL beaker containing 10 g of the TiO_2 support materials with constant stirring. The appropriate amount of manganese acetate was then added, and the mixture was stirred for 2 h. The paste obtained was dried overnight at 110°C and then calcined at 500°C for 2 h in static air. In this study, the Mn loading of the samples was 5.0 wt%.

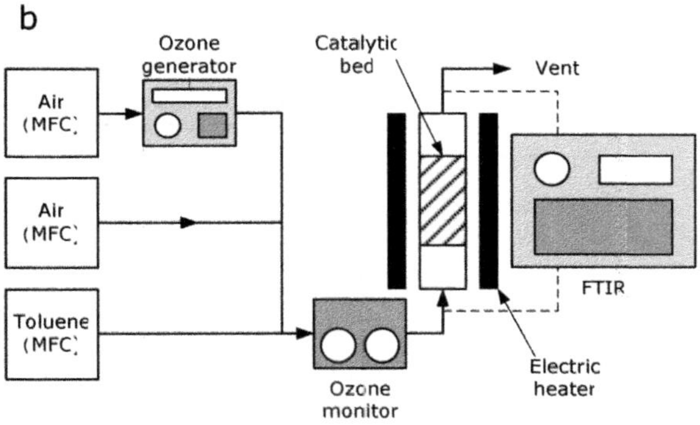

Fig. 2 Schematic diagram of the catalytic testing process

(2) Catalytic reactions

Fig. 2 shows the reaction system used in this study. The desired toluene concentration was adjusted by mixing toluene standard gas (*ca.* 200 ppm, N_2 (99.9999%) balance, Shinyang) with clean air (99.999%, Shinyang). Ozone was generated from a silent discharge ozone generator (LAB-1, OZONETECH). A set of mass flow controllers (3660, KOFLOC Co. Ltd.) were used to regulate the flow rates of the feeding gases. The catalytic bed was controlled using an electric heater, and the reaction temperature was varied from 25 to 100°C. The ozone concentrations used were 20, 50 and 100 ppm. The catalytic conversion of toluene was carried out under a gas hour space velocity (GHSV) of 20000 h^{-1}. A gas sample was analyzed using a Fourier-transform infrared spectrophotometer (MIDAC) equipped with a 10.0m optical length gas cell (volume 4.0 L). A homogeneous gaseous reaction of toluene with ozone can be neglected in this reaction system because both toluene oxidation and ozone decomposition did not proceed in the gas phase with an empty reactor. Toluene catalytic conversion and the CO_x selectivity was defined using the following equation, respectively.

3 Results and Discussions

When 50 ppm toluene was supplied to the MnO_x/TiO_2 catalyst bed, the catalyst became saturated in toluene (phase 1), reaching a steady state after 20 min. When Ozone was then added to the effluent gas (phase 2), CO and CO_2 were found.

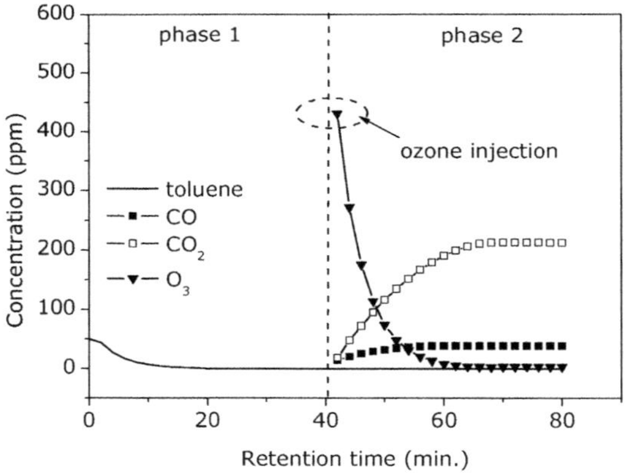

Fig. 3 Time course behavior of toluene, CO, CO_2 and O_3 with MnOx/TD-TiO_2 catalyst

During phase 1, the low residual concentration of toluene was attributed to the adsorptive properties of MnO_x/TiO_2. No reaction occurred because the formation of oxides was not detected. When the ozone was injected (phase 2), the formation of carbon oxides began. The concentrations of CO and CO_2 reached a plateau at 17 and 25 minutes, respectively. These different behaviors could be connected to different pathways [9]. Ozone as the main long-living radical was transported to the catalyst and could take part in heterogeneous oxidation reactions on its surface. Futamura et al. reported a plausible mechanism for the MnO_2-catalyzed decomposition of O_3 [5,12]. Manganese oxides are active catalysts in several oxidation or reduction reactions. Manganese oxides can be employed as catalysts for the oxidation of carbon monoxide [5,13,14].

The reaction temperature plays an important role in the catalytic oxidation of toluene with ozone. Fig. 5 shows the effect of the reaction temperature on the toluene conversion and CO_x selectivity at various temperatures in the range of 25-100°C with MnO_x/P25-TiO_2 and MnO_x/TD-TiO_2.

Toluene conversion was almost 100% at 25°C regardless of the catalysts, whereas toluene conversion decreased with increasing reaction temperature. This tendency was attributed to the adsorptive properties of MnO_x/TiO_2. Meanwhile, toluene conversion with MnO_x/TD-TiO_2 was greater than that of MnO_x/P25-TiO_2. This tendency suggests that SSA of MnO_x/TD-TiO_2 was higher than that of MnO_x/P25-TiO_2. In addition, CO_x selectivity was improved by the increase in reaction temperature. Catalytic oxidation with ozone combined with catalyst heating is effective in maintaining the catalytic activity because the heat treatment promotes the oxidation of the intermediate compounds to CO_x [6]. The differences between toluene conversion and CO_x selectivity was attributed to the desorption of formic acid from the catalyst surface to the gas phase. The differences decreased with increasing reaction temperature, suggesting that the imperfect carbon balance was gradually reduced.

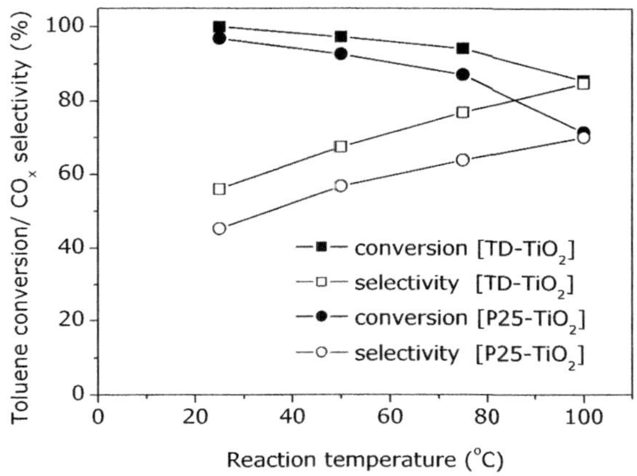

Fig. 4 Effect of the reaction temperature on toluene conversion and CO_x selectivity(MnO_x/P25-TiO_2 and MnO_x/TD-TiO_2, toluene 50 ppm; ozone conc., 500 ppm)

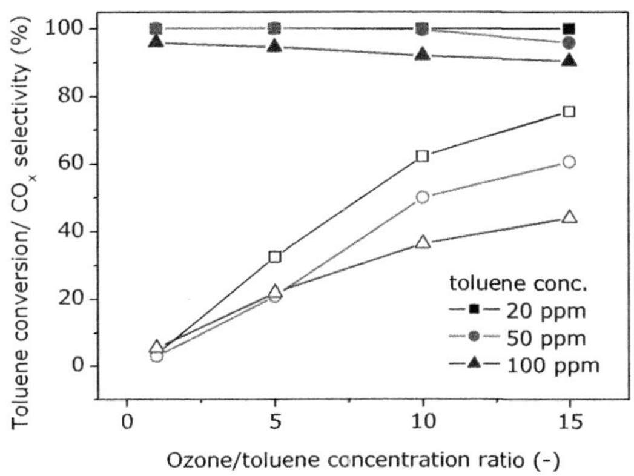

Fig. 5 Effect of toluene concentration and the ozone/ toluene ratio on toluene decomposition and CO_x selectivity (MnO_x/P25-TiO_2 and MnO_x/TD-TiO_2, solid box: toluene conversion, blank box: CO_x selectivity, reaction temperature: 25°C)

As mentioned above, ozone was transported to the catalyst and could take part in heterogeneous oxidation reactions on its surface. The ozone decomposition sequence with MnO_x/TiO_2 has been suggested to first involve the adsorption of ozone to a site to produce a free oxygen molecule and a surface oxygen atom followed by a reaction of the atom with another ozone molecule to produce two more oxygen molecules. Therefore, the higher concentration with MnO_x/TiO_2 indicates that more free oxygen molecules could be produced. Fig. 5 shows the effect of toluene concentrations and the ozone to toluene concentration ratio on toluene conversion and CO_x selectivity with MnO_x/P25-TiO_2 and MnO_x/TD-TiO_2 at 25°C. CO_x selectivity increased linearly with increasing ozone concentration. This is because the oxidation of intermediate compounds with ozone was promoted by catalyst heating. The CO_x selectivity decreased with decreasing ozone concentration due to the formation of formic acid in the gas phase. The toluene conversion increased steadily with increasing ozone to toluene concentration from 1.0 to 15.0, indicating that the reaction rate of toluene conversion on MnO_x/TiO_2 catalysts in the presence of ozone is strongly dependent on the toluene concentration. It should be noted that a lower concentration led to higher toluene conversion at the same ozone to toluene ratio. Therefore, the toluene concentration and higher ozone to toluene ratio can enhance toluene conversion.

References

[1] Drago, R.S., et al.: Appl. Catal. B-Environ. 6, 155–168 (1995)
[2] Futamura, S., et al.: Catal. Today 72, 259–265 (2002)
[3] Li, W., et al.: J. Am. Chem. Soc. 120, 9041-9046 (1998)
[4] Li, W., et al.: J. Am. Chem. Soc. 120, 9047–9052 (1998)
[5] Guo, Y.F., et al.: J. Mol. Catal A- Chem., 93–100 (2006)
[6] Einaga, H., Ogata, A.: J. Hazard. Mater. 164, 1236–1241 (2009)
[7] Einaga, H., Futamura, S.: J. Catal. 227, 304–312 (2004)
[8] Chin, S., et al.: Anal. Chem. 29, 750–754 (1997)
[9] Pratsinis, S.E.: Prog. Energ. Combust. 24, 197–219 (1998)
[10] Futamura, S., et al.: Catal. 72, 259–265 (2002)
[11] Imamura, S., et al.: J. Catal. 151, 279-284 (1995)
[12] Mirzaei, A.A., et al.: Catal. Commun. 4, 17–20 (2002)

Introduction of KNCPC (Korea National Cleaner Production Center) and Activity on Green Growth

JongHo Lee, SungDuk Kim, and JongMin Kim

Korea National Cleaner Production Center in
Korea Institute of Industrial Technology

Abstract. KNCPC is established in 1999 and have conducted many projects for improving resource productivity and expanding green management nationwide. Also, KNCPC complies with global environmental regulation. Moreover, KNCPC has prepared national key R&D projects and provided web service, cleaner production D/B.

1 Introduction

KNCPC is a non-profit organization established in 1999 under the support of the Korean Ministry of Knowledge Economy (MKE). We became a member of UNIDO/UNEP NCPC Network in 2001; hosted and conducted lots of international symposiums and projects so far. Our major roles are to develop and manage projects, comply with global environmental regulations, and plan national key R&D projects.

KNCPC organization consists of 4 offices as the fig.1.

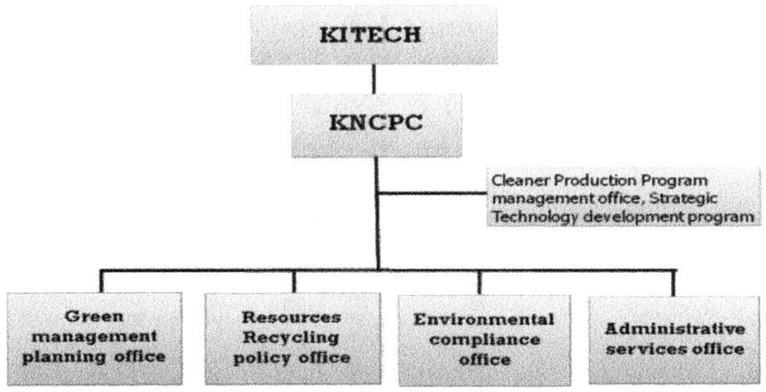

Fig. 1 KNCPC organization

2 History of KNCPC

KNCPC has conducted planning and managing projects and hosted symposiums science 1999. The table 1 shows the history.

M.-W. Han and J. Lee (Eds.): EKC 2010, SPPHY 138, pp. 35–40.
springerlink.com

Table 1 History of KNCPC

Year	Achievement
1995	Enactment of "Legislation Promoting the Shift to Environment-Friendly Industrial Structure"
1995	CP Technology Development Program" initiated by MOCIE (MKE)
1999	Establishment of KNCPC
2001	KNCPC joined UNIDO/UNEP NCPC Network
2003	KNCPC hosted 2nd Asia-Pacific Regional Expert Meeting in Sustainable Consumption & Production
2003	KNCPC launched Green Partnership project
2005	KNCPC planed Eco-Industrial Park(EIP) project
2006	KNCPC started Remanufacturing project
2007	KNCPC initiated Regional Eco-innovation project
2007	REACH Business Service Center was established
2008	KNCPC planed New Growth Engine for Energy and Environmental division
2009	KNCPC hosted International Green business Symposium

3 Improving Resource Productivity

For improving resource productivity, we have conducted "Urban Mine project" which is to collect-separate-select-smelt-refine valuable metals from wastes (Fig. 2).

"Re-Manufacturing project" which is to collect- disassemble-clean-inspect-repair the end-of-life products; KNCPC signed MOU with National Remanufacturing Center of US in 2004 developing remanufacturing technologies for automobile transmission and a toner cartridge(Fig. 3).

"Eco-Industrial Park" which is industrial complex increasing resource efficiency and minimizes pollution through exchanging the waste; The pilot program began in 5 major industrial complexes (Ulsan, Yeosu, Sihwa, Pohang, Chungju) in 2005 and, now, the annual profit of $23M produced through 7 projects commercialized with reduction of 100K ton of CO2 per year(Fig. 4).

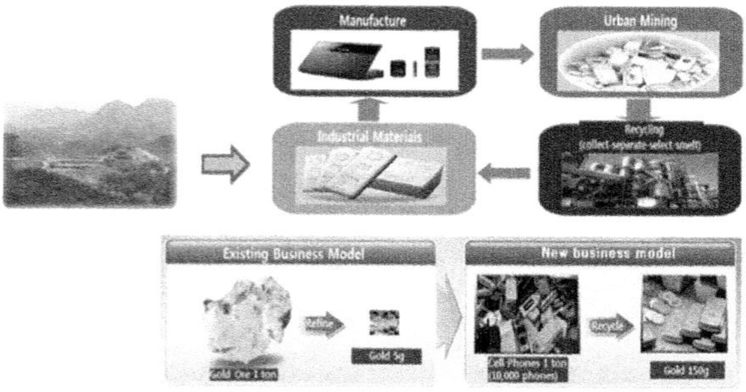

Fig. 2 Urban Mine Project

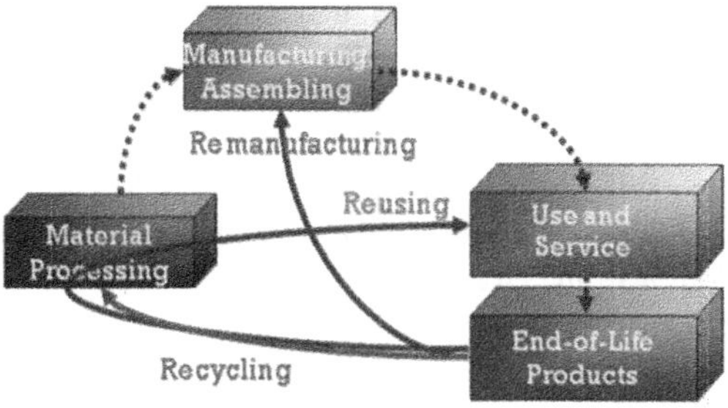

Fig. 3 Re-Manufacturing Project

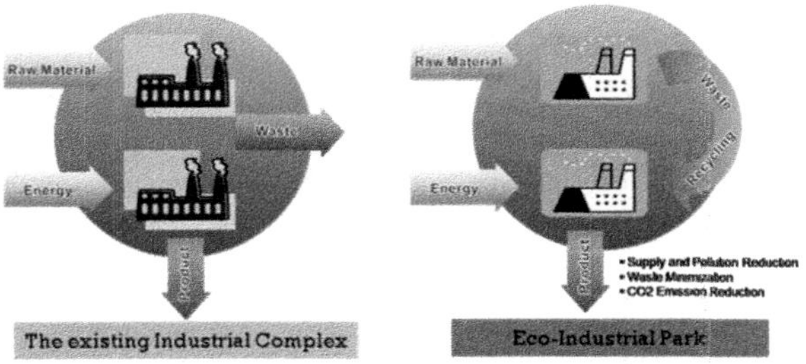

Fig. 4 Eco-Industrial Park Project

4 Expanding Green Management Nationwide

For expanding green management nationwide, we have conducted "Green Partnership project" which is enterprise's cooperation in its supply chain to achieve economic and environmental profits through sharing knowhow, management system and technology related to the environment(Fig. 5).

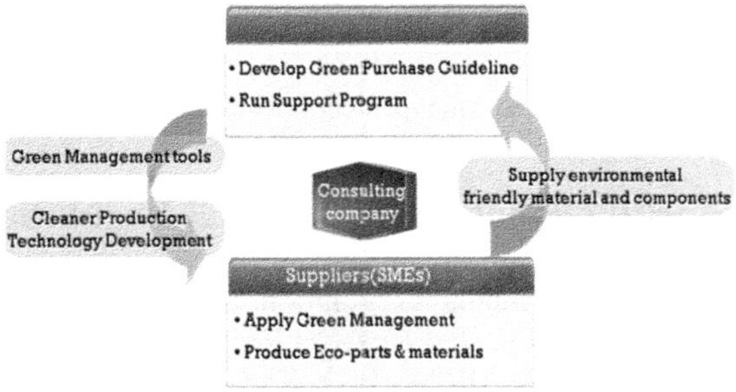

Fig. 5 Green Partnership Project

"Regional Eco-Innovation project" which is cooperation-based on regional authorities, businesses and consultants through consulting companies and networking between regional businesses; The project has been conducted in 5 megalopolis(Pusan, Inchon, Taegu, Ulsan, Taejon) in 2007(Fig. 6).

"Chemical Management Service project" which is CMS provider manages all processes from purchasing and usage to disposal of chemicals treated at business sites; KNCPC has demonstrated for automaker and electronic industrial Part since 2008(Fig. 7).

Fig. 6 Regional Eco-Innovation Project

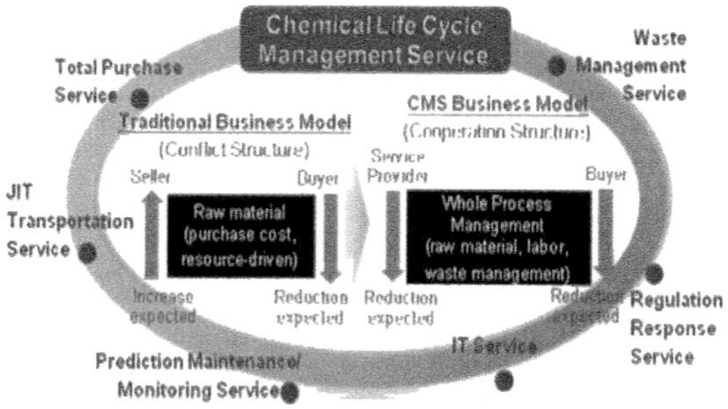

Fig. 7 Chemical Management Service(CMS) Project

5 Complying with Global Environmental Regulation

To comply with global environmental regulation, KNCPC made REACH Business Service Center('09. May) and hosted REACH Expo 9th(08, Feb), and joint promotion team for responding REACH(08. May). KNCPC also has promotion plans to establish global environmental regulation on tracking D/B and to promote consulting part(Fig. 8).

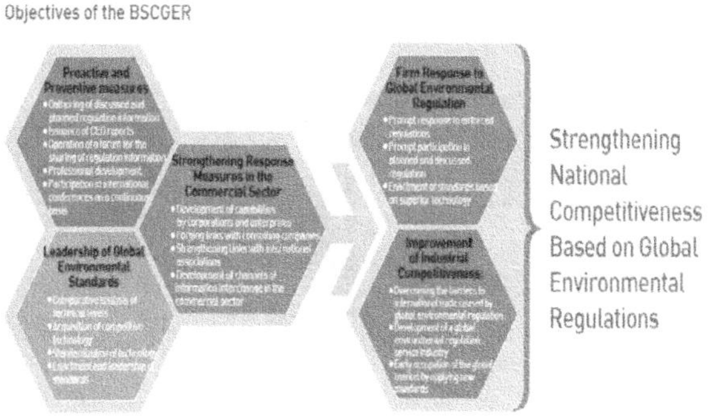

Fig. 8 Objectives of the BSCGER(Business Service Center for Global Environmental Regulation)

6 Planning and Managing National Key R&D Projects

KNCPC has prepared national key R&D projects likewise "New Growth Engine in Energy & Environments Part" for president report (performing role of energy & environments part assistance administrator, "Green Ocean 100 Specific Technology Plans" to improve nation the quality of life and solve global worming problem and depletion of natural resources, and "Green Growth Plan for Development of Industry through Knowledge Innovation" to achievement of green conversion value chain.

"Green Ocean 100 Specific Technology Plans" in 3 key R&D projects are divided into 9 major areas: human wellbeing space, safety water, zero waste plus resource, CO_2 reuse and resourcing, low carbon material and process, green solution by IT, clean coal, biomass economy, sustainable product service. KNCPC has planed and managed them and fig. 9 and 10 show the briefing and a goal of one of them.

Fig. 9 Briefing on "Green Ocean 100 Specific Technology Plans"

Fig. 10 Goal of low carbon material and process

7 Providing Cleaner Production D/B (CP-NET)

KNCPC has provided "CP-NET" by web that is total D/B on cleaner production (Fig. 11).

Fig. 11 CP-NET

8 Cooperation between KNCPC and EKC

Expected cooperation would be;

- Cooperating joint research related to technologies and policies on green growth.
- Co-hosting a forum, workshop, to make a critical review and share information.
- Supporting partner matching, coordination, subjective information gathering on experts, laboratories, institutes, companies in EU.

Art, Science and the Public: Focusing on Science as a Resource for Creative Art

Seonghee Kim[1,2,*]

[1] Korean Cultural Studies, School of Arts and Humanitties, The Nottingham Trent University, UK
[2] Centre for Effective Learning in Science, School of Science and Technology, The Nottingham Trent University, UK
seongheekim69@hotmail.com

Abstract. Art's search for new subjects and methods and science's need for effective communication have led to the creation of what is known as Sci-Art. It is the central argument of this paper that collaboration between creative and scientific disciplines can play a useful role in society, but that this potential is held back by misunderstanding of the roles of art and science. The main purpose of this study is to determine the relationship between artists and scientists, focusing on the visualisation of DNA. This study identified their shared approaches to its representation, and explore the history of DNA as an iconic form. An additional purpose of this study is to analyse the importance of the role of collaboration between scientists and artists including its application to education. My method is to review Sci-Art work and analyze the benefit of collaboration between science and art. Part of this research will focus on the benefits of Sci-Art collaboration for education. Collaborative artworks and exhibitions are the final outcome of this project; they explore the ways in which Sci-Art can be developed as a useful form of interdisciplinary practice. These creative methods provide a route to a deeper understanding of the relationship between art and science. This study demonstrates through a combination of theoretical argument and creative practice that Sci-Art has the potential to: Act as an aid to understanding difficult scientific concepts; add to debate about the ethical issues surrounding science; and increase the effectiveness of education.

1 Introduction

Since 1990, an increasing number of artists have been inspired by biotechnology and the social and moral issues that surround it. Sci-Art is the generic name given to artworks that uses scientific concepts, images or technology. Firstly, I would like to discuss the relationships between art and science. As Martin Kemp [1] pointed out, to generalise about the relationship is not so much hazardous as impossible because neither science nor art are homogeneous categories and they both deal with a number of aspects: Science ranges from the observed complexities of environmental biology to the unseeable dimensions of theoretical physics; Art extends from the figurative representations of nature to the elusive abstractions of

* Corresponding author.

M.-W. Han and J. Lee (Eds.): EKC 2010, SPPHY 138, pp. 41–45.
springerlink.com

conceptual art. Even if we take one subject, molecular biology, for example – or in particular, DNA – its expansion into art may vary from iconographical reference to reflection of biotechnology on our future.In this aspect, I believe that both art and science should be considered as a very powerful 'engine' in our culture. Thus, it could be argued that we still need to research in terms of interdisciplinary aspects.

2 Science, Art and Education

Sci-Art is a relatively new and exciting area, bringing artists and scientists together to work on ideas that have common themes or problems. Thus several foundations and organisations promote many such partnerships and projects including artists-in-residence schemes. I will argue that Sci-Art can promote public engagement with science and raise important questions about how science affects our world.

Sci-Art emerged from a socio-cultural context that has provided major support to collaborations between artists and scientists since at least 1996, when the Wellcome Trust launched their Sci-Art scheme. In 1998, the U.K. National Endowment for Science, Technology and the Arts was created, and in 1999 the Sci-Art Consortium was established, comprising no fewer than five major funding bodies coming together to support art and science projects. At the end of 2001, the Arts Council England and Arts and Humanities Research Council agreed to set up a new joint funding strand for interdisciplinary research fellows working across science, technology and art [2,3].

Ken Arnold, head of the exhibitions department at the Wellcome Trust, accurately articulated the necessity of these projects; firstly, science has provided some artists with inspiration in the areas of medium, message and location; secondly, the world of science can gain in many different ways from the arts, and bring new perspectives and insights; finally, the true power of science and art as a union is that the union can build a more engaged relationship with the pubic, not the practical breakthroughs [4].

C.P. Snow argues that students majoring in science should study art subjects, and *vice versa*.

> **This study could be grafted into any of our educational systems, at high school or college levels, without artificiality and without strain. I dare say that, as usual, this is an idea which is already floating around the world, and that, as I write this paragraph, some American college has already laid on the first course [5].**

Snow explained that some of American colleges already started this type of interdisciplinary study by 1960. He claimed that the two cultures have to meet each other in an early education period (secondary college level) to avoid creating a gap between them. Since Snow pointed out the importance of crossover education 40 years ago, some educators have found that many students have difficulties with and are less interested in science subjects which have separated from other subjects. They share Snow's opinion that art and science should be interdisciplinary learning subjects.

3 Workshop Using Science as a Resource for Creative Art

It is commonly considered that the art and science are on opposite ends of educational spectrum because similarities and differences between two academic disciplines seem to be misunderstood and at variance with each other. However, it represents social and moral issues to the public, and many artists express their concepts of science in the post-genome era. In this regard, the consideration of the relationship between art and science is necessary for potential roles in art. In particular, most interdisciplinary education in art and science are carried out in

Fig. 1 Final art work of 3-D sculpture using laboratory ware and natural materials

science-based schools or institutes, focusing effects or influences of art on science such as inspiration. However, science is not considered as a subject for creative art *per se*. Thus the aim of this workshop is to provide another aspect and understanding of science for art, particularly focussed on DNA and Biotechnology. The art works with students are presented in figure 1.

4 Conclusion

E. O. Wilson [6] indicated in his book, *Consilience: The Unity of Knowledge*, that the gap described by C. P. Snow between two cultures of the sciences and the humanities, including art, continues to exist today. For these reasons, I am convinced that to gain a better understanding of one subject, two different disciplines should be learned at the same time in school, comparing their similarities, differences, and interactions.

As shown in Figure 2, science, art, and education have not only their own domains in our culture where human activities are reflected, but also intersecting domains. I would like to close by proposing that Sci-Art projects can be adopted as a new bridging discipline in education to improve communication between art and science, and help to understand their intrinsic and extrinsic conceptions.

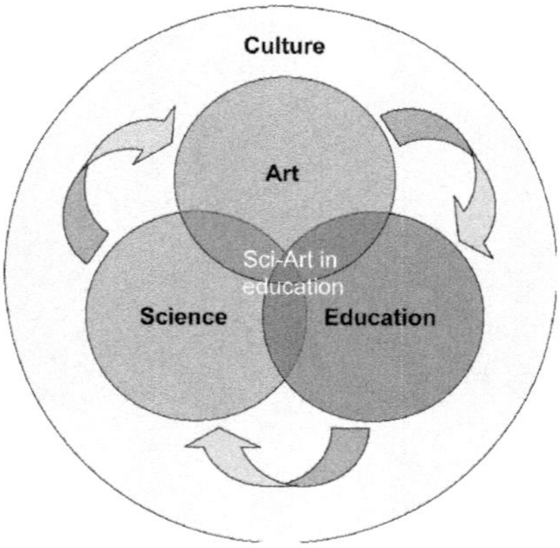

Fig. 2 The necessity of collaboration between science and art in education

I would like to introduce my artwork which present how science can be creative, beautiful and inspiration as well as explore our lives and culture through the medium of sculpture (Fig.3).

Fig. 3 Life is Endless Desire. Exhibited as part of the Korean Science Festival 2005 from 12th to 21st of August 2005 in Daejon, South Korea.

References

[1] Kemp, M.: From science in art to the art of science. Nature 434, 308–309 (2005)
[2] Ferran, B., Leach, J., White, T.: Arts and Science Research Fellowships: Art Council England and Arts and Humanities Research Board. Leonardo. 39(5), 441–445 (2006)
[3] British Council, Art and science: a new partnership [online]. Briefing sheet 23, British Council (2003), http://www.britishcouncil.org/science-briefing-sheet-23-sci-art-jun03.doc (accessed July 7, 2006)
[4] Arnold, K.: Between Explanation and Inspiration: Images in Science. In: Ede, S. (ed.) Strange and Charmed, 1st edn. Calouste Gulbenkian Foundation, London (2000)
[5] Snow, C.P.: The Two Cultures: And A Second Look. Cambridge, New York (1963)
[6] Wilson, E.: Consilience: The Unity of Knowledge, 1st edn. Alfred A. Knopf, Inc., New York (1999)

Adaptive Strategy Decision Mechanism for StarCraft AI

Sangho Yi*

INRIA Grenoble Rhone-Alpes, France
sangho.yi@inria.fr

Abstract. Recent advancement of computer systems and their global networks have enabled us to enjoy many kinds of on-line games. StarCraft has been the one of the well-known on-line strategy simulation games, since it was made in 12 years ago. The game users have been extremely improved their playing level, but the AI (artificial intelligence) system of the game was rarely updated. At the consequence, current AI level is much poorer than that of the human players. In this paper, we present a noble AI mechanism for StarCraft based on the adaptive strategy decision mechanism. We implemented the proposed mechanism on the real StarCraft: BroodWar binary code. We compared the performance of the modified AI code, and the results show that the new AI code outperforms the existing artificial intelligence of the game.

1 Introduction

In these days, video game industry has been increased significantly, and now it has a major impact on the world economics. Especially, the computer games market has grown dramatically over the last decade by human demands. Current technologies enable the computer games to connect all the people at anytime, and anywhere in the world.

Based on the high-speed networks, distributed systems, and significant computing power, they become not only short-term enjoying entertainment, but also a kind of long-term sports whole the world. For example, the first e-Sports competition, *World Cyber Games* has been held since 2001 [1], and now it is the biggest annual Game-Olympiad in the world. Also, the *World e-Sports Games* [2] has been held annually after 2004. In addition, the words '*professional players*' are not exclusively possessed by the general sports. We already know the existence of the '*professional-gamers*' on many popular computer games [3, 4]. StarCraft [5] was made by Blizzard Entertainment in 1998, and it has been the most famous real-time strategy computer game in the world.

Even if above e-Sports have significant amount of players and fans, even if they are so popular in this time, they still have several challenges because of their characteristics. First, players may need to play with computer AI systems. Even if

* Corresponding author.

M.-W. Han and J. Lee (Eds.): EKC 2010, SPPHY 138, pp. 47–57.
springerlink.com

player vs. player games, the players may use some 'macro' commands which should be run with computer's AI. Second, someone (vendors, players or third-parties) should manage, update, and maintain the game core framework, because this is based on the large-scale computer software systems. In those points of view, we cannot assure the longevity and quality of the e-Sports, for example, StarCraft was made 12 years ago, and the main core was rarely improved after the year 1998. Just a few patches were distributed to fix bugs on it. At the consequence, level of the artificial intelligence becomes relatively much poorer than that of the human players. The players may win the game even if one human player fights with multiple computer AIs on the same custom game. Thus, StarCraft AI code has to be upgraded and patched to relieve huge gap between computers and players.

Considerable efforts [6-13] have been made to improve the poor AI of the StarCraft. However, some of them used only '*cheat codes*' to improve the power of AI in the order of magnitude [6]. That kind of work is just a piece of cheats, and they do not have any improvement in quality compared to the existing AI. In this paper, we present a noble AI mechanism for StarCraft based on the adaptive strategy decision mechanism. It was designed and developed to improve current level of AI. We implemented the proposed AI mechanism on the real StarCraft binary code, and made several experiments. The results show that the performance of the proposed AI mechanism outperforms the current AI on the StarCraft.

The rest of this paper is organized as follows. Section 2 presents work related to AI development on the video games. Section 3 presents the design and implementation of a noble AI mechanism on the StarCraft in detail. Section 4 evaluates the performance of the proposed AI mechanism compared with the current AI on the StarCraft. Finally some conclusions and possible future work are given in Section 6.

2 Related Work

In this section, we present a brief overview of previous research efforts on the AI of the video games and its implementation [6-15].

In 2001, IBM announced Robocode contest for having both educational effect and better AI for mobile tank robots [14]. This has been well-known contest in the world, and many algorithms have been made on top of the Robocode's framework.

In [15], Ian and James proposed a script language for programming artificial intelligence of games on Sony's PlayStation 2. This is a rule-based object-oriented programming language which called by *RC++*. It was designed to alleviate the complexity of programming codes in *C* language, and to give more flexible representative power. By using the *RC++*, the programmers may write out more efficient and well-structured codes of artificial intelligence for the games on PlayStation 2.

In addition, in [8], ScAIEdit (StarCraft Artificial Intelligence Editor) was proposed to represent and program the AI code of the StarCraft. ScAIEdit is a set of converter and editor of the artificial intelligence on this game. By using it, programmers can easily retrieve and modify the original source code of the StarCraft. It represents the machine level code by their own language. In compiling phase, it transforms the source code to the binary code of the StarCraft.

Also, there are considerable efforts [6, 16, 17] on developing and improving the artificial intelligence of the StarCraft.

First of all, an unofficial AI patches of the StarCraft were made by some anonymous programmers [6]. However, they used only '*cheat code*' to dramatically improve the existing artificial intelligence of the StarCraft. Thus, it is an unfair improvement to compare with the existing one.

In several players' forums, bunch of strategies have been developed based on the cost and time analysis under constraints of the required time and resources. However, no strategy did not be the greatest one, because StarCraft is a multi-player game, and there is no optimal solution. The strategies of each player may affect one another, and unlike to Chess or Game of Go, StarCraft players do not share their units, the amount of harvested resources, and etc. Therefore, to make an efficient and robust StarCraft AI, we have to adapt itself with dynamic conditions.

Recently (in 2009), Blizzard Entertainment opened the StarCraft:BroodWar API [9], which is C/C++ based programming interface to the computer AI robot. They also announced a AI programming contest [10] to get more volunteers on this opening.

Since the beginning of 2010, StarCraft II has been tested with volunteers, and the AI programming method [11] was discovered by anonymous users, and they developed several basic AI mechanisms to play with human players.

3 Design and Implementation of Adaptive Strategy Decision Mechanism on the Starcraft

In this section, we present design of a noble StarCraft AI mechanism which is based on the adaptive strategy decision mechanism. Also, we show programming method in detail to implement the AI mechanism on the StarCraft.

3.1 Starcraft in a Nutshell

In the world of the StarCraft, there are three kinds of races: Terran, Zerg, and Protoss. The three races have their own characteristics on buildings, units, and technical or magical skills. Buildings are used to train units and research some skills, and the units gather resources and construct some buildings, and fight with the opponents. Resources are statically deployed on the field, and the gathered resources are used to train units or construct some buildings with its production costs. Then, the trained units are organized into several troops, and it can attack to the opponents or defense them. The mission of the game is killing all other opponents, and thus, the efficiency of gathering resources, training units, constructing buildings, and using better military strategy are very important in the game.

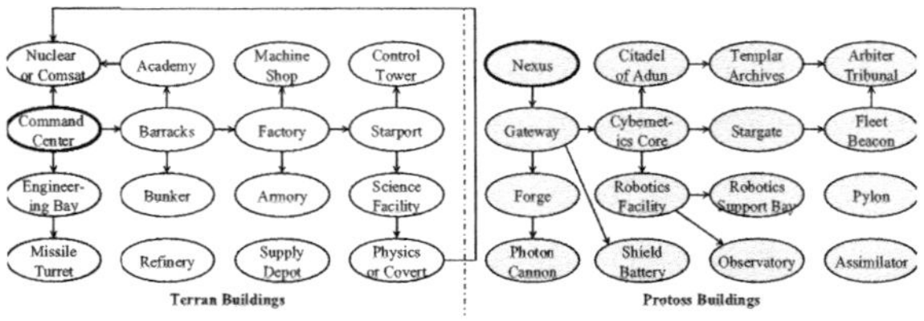

Fig. 1 Dependency relationships between buildings in Terran and Protoss races

In StarCraft, all buildings have their own dependency relationships. Fig. 1 shows the dependencies of the Terran and Protoss races [5]. For example, if we have no Barracks, we cannot build a Factory, and thus, we must build a Barracks first when we have to build a Factory. Also, we need to consider an appropriate number of peasants. Too many peasants are wasteful, but too small also be a cause of deficiency of the gathered resources.

3.2 Adaptive Strategy Decision on Starcraft

In real-time strategy games, strength of a player is determined by efficiency of the used strategies. The strategies on these kinds of games can be divided into following items: build-orders, combination of units, attack and defense timing, micro/macro unit control, and combat tactics. In those, current StarCraft AI is so weak in all the aspects compared with human players. That's because the computer's AI does not have the appropriate strategy decision method, and it was rarely improved since 1998. In the result, the current AI is not competitive compared with the players.

We have done related work on developing StarCraft AI mechanisms. In [12], we presented how to program StarCraft AI codes on the game, and in [13], we presented a knowledge-based strategy decision mechanism for improving StarCraft AI. In the previous work, we implemented the prototype version, and we shows significant performance improvement compared with the existing AI.

However, the modified AI was so predictable, and straight forward, because at that time we had only static decision mechanism which do not consider the dynamic situations. It did not work well with human players when the players understand their static decision functions.

Thus, in this paper, we propose an adaptive strategy decision mechanism to improve the existing intelligence of the StarCraft. The mechanism enables that each player adaptively decides an appropriate strategy by using the recent information from the opponents. The information consists of recent build-orders, combination of units, and timing of attack and defense. In addition, computer players does not need to scout the enemies because they already know what the enemies do. Thus, it easily enables the adaptive strategy decision mechanism on the computer's artificial intelligence.

Some dividable stages exist on the StarCraft. Based on many experiences from users [16, 17], the whole game has been considered as the three stages as follows.

- *Opening*: This is the first few minutes of the game. Similar to the 'opening stage' in Chess and Game of Go, we have very well-known build-orders. In this stage, selecting an appropriate build-order based on the opponent's strategy is very important because it determines the successful defense from the early attacks from the opponents.
- *Middlegame*: This is started by the first massive attack. In this stage, players are preparing larger combat by expanding and increasing their own troops and production power. Thus, the utilization of the resources (gathering resources and producing units) is the most important in this phase.
- *Endgame (Last stage)*: This is after one become the dominant player in the game. Normally it comes with some massive attack. In this, a series of massive attack may occur, and the players may use almost all the field of the game to gather resources. Therefore, selecting an efficient massive attack and defense timing is the most important in this phase.

Based on the divided stages, we designed the adaptive strategy decision mechanism based on the basic principles listed below.

- *Use well-known strategies / build-orders.*
- *Utilize received information from opponents.*
- *Utilize known characteristics of units.*
- *Exploit computer's macro commands.*
- *Have various strategies to prevent prediction.*
- *Do not save resources for the future use.*

Table 1 Notations and functions used in this paper

Notation	Description
s_i	i-th possible strategy
t_c	Current time
t_e	Required execution time of the opponent's expected strategy
t_l	Laxity time from t_c to t_e
$t_r(s_i)$	Required execution time for s_i
$d(s_i)$	Degree of dangerousness of using s_i
d_{thres}	Predefined maximum threshold value of the degree of dangerousness

Table 1 shows the notations and functions used in this paper. The s_i represents i-th possible strategy of a computer player. Thus, the computer player has multiple strategies (i = 0, 1, … , n). The strategies include both attack and defense purposes. The t_e is the execution time of the opponent's expected strategy. The t_l is the remaining time to the expected execution time opponent's strategy. The $t_r(s_i)$ is the required time to execute the strategy s_i, and the $d(s_i)$ represents the degree of dangerousness of executing the strategy s_i. Finally the d_{thres} is the predefined maximum threshold value of the degree of dangerousness.

Algorithm 1. Adaptive Strategy Decision

- do the following when a computer player needs to select the next strategy:

$k := 1, x := 1$
calculate t_e
$t_l := t_e - t_c$
for all s_i $(i=1,2,\ldots,n)$ **do**
  if $t_r(s_i) < t_l$ **then**
    $s_{ten-k} := s_i$
    increase k
  end if
end for
for all s_{ten-i} $(i=1,2,\ldots,k)$ **do**
  if $d(s_{ten-i}) < d_{thres}$ **then**
    $d_{select-x} := d(s_{ten-i})$
    $s_{select-x} := s_{ten-i}$
    increase x
  end if
end for
  randomly select s_{final} **from** $s_{select-i}$ **where** $s_{select-i}$ $(i=1,2,\ldots,x)$

Algorithm 1 shows the adaptive strategy decision mechanism in detail. First of all, Algorithm 1 checks $t_r(s_i) < t_l$ for all possible strategies. Then, it saves the tentative strategies in s_{ten-i} $(i=1,2,\ldots,k)$. Second, it checks the degree of

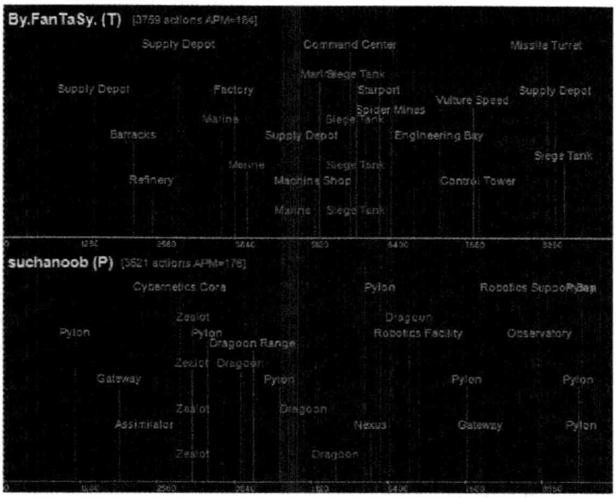

Fig. 2 An example of build-orders of a game

dangerousness of all possible s_{ten-i} . If the $d\left(s_{ten-i}\right)$ is smaller than d_{thres} , the strategy is saved on $s_{select-x}$. Finally, Algorithm 1 randomly selects one strategy s_{final} from the all possible strategies on $s_{select-i}\ (i=1,2,\ldots,x)$.

The major challenge of using Algorithm 1 comes from measuring required time for each strategy. To do this, we used BWChart Replay Analyzer [18] to monitor the build-orders and the execution time for each strategy from the previous replay records. Figure 2 shows an example of build-orders of two players. By using the replay records of the top-class amateur players, we found the reasonable $t_r\left(s_i\right)$ for many possible strategies.

3.3 Implementation

To implement the adaptive strategy decision mechanism on the StarCraft, we used ScAIEdit [8] and WinMPQ [19] tools. In [12], we have summarized the overall programming methods of the StarCraft AI with many kinds of practical programming examples. In the system of the StarCraft, the AI code consists of two-level modules. The first one is on the core routines of the binary executable program, and modifying the core routines is relatively hard. Even we can use the recently opened BroodWar API, there is lack of documents and resources for doing this. This is not the main focus of this paper. The second one is separated from the binary program. They reside in the *Patch_rt.mpq* file of the StarCraft folder, and the code is easy to program in script language level by using the ScAIEdit and WinMPQ tools.

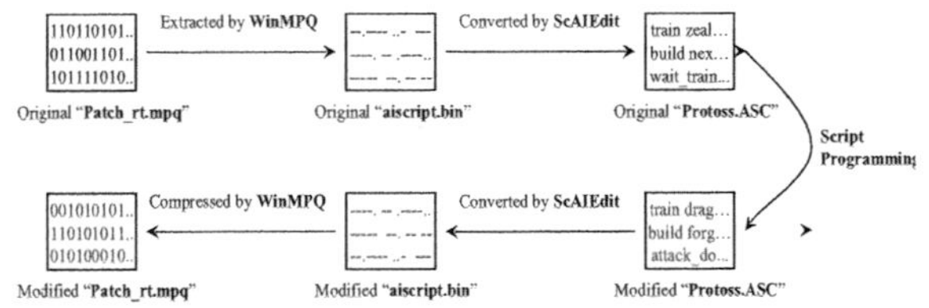

Fig. 3 Basic flow of developing StarCraft AI

Figure 3 briefly shows the overall flowchart of the AI programming method on the StarCraft. In this work, we extracted the AI source code from the original *Patch_rt.mpq* file, and then, we implemented the adaptive strategy decision mechanism on top of the script code[1].

4 Performance Evaluation

In this section, we evaluate the performance of the proposed artificial intelligence based on the adaptive strategy decision mechanism.

To fully evaluate the real performance of the modified artificial intelligence of the proposed artificial intelligence, we did experiments on the StarCraft. We used StarCraft: BroodWar ver. 1.15.2 [5] game package[2], and the series of official map files from [21].

We made experiments in both micro and macro scopes. Firstly we had micro benchmarks when a computer plays a Terran race while the opponent plays a Zerg race with fast 6 Zerglings rushes (the fastest attack strategy in StarCraft). We used the official map, 'The Hunters KPGA', and have tested if the computer Terran can defense the fast attack without significant sacrifice, and the Table 2 shows the results. In the results, the proposed AI outperforms the current AI on StarCraft, and the proposed AI can defense fast Zergling attacks for most cases except for the case of using only 4 Drones[3].

We have tested another popular initial attack situations, player's fast Zealot rushes to the computer AI player. We used the same condition for the rest, and Table 3 shows the results when computer plays a Terran race.

Secondly, we made macro benchmarks (whole game). We implemented the new AI mechanism on the Terran races, and we tested them with the unmodified AI. In the matches, we used the 'Python' map [21], and Tables 4 and 5 show the results of 1 vs 1 matches.

---

[1] The modified AI patch file is available at the website [20].

[2] The same AI patch can work with the recent version (1.16.1) of StarCraft: BroodWar.

[3] When a player uses a 4 Drones, 6 Zergling rush strategy, the opponent's troops can reach to the computer's base within 2 minutes and 10~40 seconds. If Terran builds a Bunker, it takes about 2 minutes and 30 seconds. Thus in some cases, it's almost impossible to defense the fastest Zergling rush.

Table 2 Experimental results of defensing fast Zergling rushes

Build-order	Proposed AI (Terran race)	Existing AI (Terran race)
4 Drones, 6 Zergling rush	3 / 10 (30% of defense)	0 / 10 (0% of defense)
5 Drones, 6 Zergling rush	5 / 10 (50% of defense)	0 / 10 (0% of defense)
7 Drones, 6 Zergling rush	9 / 10 (90% of defense)	2 / 10 (20% of defense)
9 Drones, 6 Zergling rush	10 / 10 (100% of defense)	4 / 10 (40% of defense)

Table 3 Experimental results of defensing fastest Zealot rushes

Build-order	Proposed AI (Terran race)	Existing AI (Terran race)
2 Gateways, 3 Zealots	8 / 10 (80% of defense)	2 / 10 (20% of defense)
3 Gateways, 6 Zealots	10 / 10 (100% of defense)	4 / 10 (40% of defense)
3 Gateways, 12 Zealots	10 / 10 (100% of defense)	7 / 10 (70% of defense)

Table 4 Experimental results of '1 vs. 1' matches against the unmodified Zerg computer AI

Build-order	Proposed AI (Terran race)	Existing AI (Terran race)
Average playing time	15m 21s	19m 41s
Wins / losses	20 / 0	11 / 9
Average resource use	14085.5	11499.2

Table 5 Experimental results of '1 vs. 1' matches against the unmodified Protoss computer AI

Build-order	Proposed AI (Terran race)	Existing AI (Terran race)
Average playing time	15m 17s	19m 51s
Wins / losses	20 / 0	8 / 12
Average resource use	14985.5	13903.5

Based on the results, we observe that the proposed mechanism outperforms the previous AI mechanism on StarCraft in terms of winning ratio, the amount of harvested resources, and the total playing time (less is better for the playing time).

5 Conclusions

In this paper, we proposed a noble AI mechanism for StarCraft based on the adaptive strategy decision mechanism. It was designed and developed to improve current artificial intelligence of the StarCraft. We implemented our mechanism on the real binary code, and we have tested it with the previous computer AI mechanism. Our micro and macro benchmarks showed that the performance of the proposed AI mechanism outperformed the existing AI on the recent version of tarCraft.

6 Future Work

We are currently extending our work to design and implement more efficient and adaptive strategy decision mechanism on the StarCraft, and we will draw the level of the artificial intelligence with user's level. In another study, we have implemented the proposed mechanism to play with with 2 previous AI machines (1 vs. 2), but then, the proposed AI worked in very defensive, and did not show aggressive attack patterns. We will extend our study by combining the current mechanism with BroodWar API [9] to have more possible way of implementing the adaptive strategy decision mechanism.

7 Availability

Data used in this study, the source code and binary and some results are available under the following web URL: http://aistarcraft.sourceforge.net

References

[1] World Cyber Games, http://www.worldcybergames.com (web)
[2] World e-Sports Games, http://www.worldesportsgames.com (web)
[3] KeSPA(Korean e-Sport Associaton) ProGamers, http://www.progamer.or.kr (web)
[4] Major League Gaming, http://www.mlgpro.com (web)
[5] StarCraft:BroodWar, http://www.battle.net/scc (web)
[6] BroodWar Artificial Intelligence Project, http://www.entropyzero.org/broodwarai.html (web)
[7] StarCraft Campaign Creations, http://www.campaigncreations.org (web)
[8] ScAIEdit Tool, http://www.camsys.org (web)
[9] BWAPI - An API for interacting with Starcraft: Broodwar, http://code.google.com/p/bwapi/ (web)
[10] StarCraft: BroodWar AI Competition, http://eis.ucsc.edu/StarCraftAICompetition (web)
[11] Writing your own StarCraft II AI, http://sc2.nibbits.com/articles/view/10/writing-your-own-starcraft-ii-ai (web)
[12] Yi, S., Heo, J., Cho, Y., Hong, J.: Programming method for improving performance of artificial intelligence on StarCraft. In: Proceedings of 2006 Winter Korea Game Society Conference, pp. 141–146 (2006)
[13] Yi, S., Heo, J., Cho, Y., Hong, J.: Adaptive Reasoning Mechanism with Uncertain Knowledge for Improving Performance of Artificial Intelligence in StarCraft. Journal of The Korean Society for Computer Game 7, 19–32 (2005)
[14] Hartness, K.: Robocode: Using Games to Teach Artificial Intelligence. Journal of Computing Sciences in Colleges 19(4), 287–291 (2004)
[15] Wright, I., Marshall, J.: RC++: a rule-based language for game AI. In: Proceedings of First International Conference on Intelligent Games and Simulation (2000)
[16] YGosu game forum, http://www.ygosu.com (web)
[17] Unofficial ProGamer Ranking, http://pgr21.com (web)
[18] BWChart Replay Analyzer, http://bwchart.teamliquid.net/us/bwchart.php (web)
[19] WinMPQ, http://shadowflare.gameproc.com (web)
[20] StarCraft:BroodWar Artificial Intelligence Patch, http://aistarcraft.sourceforge.net (web)
[21] Official Competition Maps for StarCraft: BroodWar, http://www.mapdori.com (web)

Nonleptonic Two-Body Decays of the B_c Mesons in Light-Front Quark Model

Ho-Meoyng Choi[1,*] and Chueng-Ryong Ji[2]

[1] Department of Physics, Teachers College, Kyungpook National University, Daegu, Korea
homyoung@knu.ac.kr
[2] Department of Physics, North Carolina State University, Raleigh, NC, USA

Abstract. We study exclusive nonleptonic two-body $B_c \to (D_{(s)}, \eta_c, B_{(s)}) + F$ decays with F (pseudoscalar or vector meson) factored out in the QCD factorization approach. The nonleptonic decay amplitudes are related to the product of meson decay constants and the form factors for semileptonic B_c decays. As inputs in obtaining the branching ratios for a large set of nonleptonic B_c decays, we use the weak form factors for the semileptonic $B_c \to (D_{(s)}, \eta_c, B_{(s)})$ decays in the whole kinematical region and the unmeasured meson decay constants obtained from our previous light-front quark model.

1 Introduction

The discovery of the B_c meson by the Collider Detector at Fermilab(CDF) Collaboration[1] and the subsequent measurement of its lifetime have provided a new window for the analysis of the heavy-quark dynamics and thus for an important test of quantum chromodynamics. Furthermore, the Large-Hadron-Collider(LHC) is expected to produce around ~5 x 10^{10} B_c events per year[2]. This will provide more detailed information on the decay properties of the B_c meson. Since the B_c mesons carry flavor explicitly (b and c) and cannot annihilate into gluons, they are stable against strong and electromagnetic annihilation processes. The decays of the B_c meson are therefore only via weak interactions, which can be divided into three classes at the quark level: (1) the $b \to q$ ($q = c,u$) transition with the c quark being a spectator, (2) the $c \to q$ ($q = s,d$) transition with the b quark being a spectator, and (3) the weak annihilation channels. In fact, both b-and c-quark decay processes contribute on a comparable footing to the B_c decay width.

Because the b and c quarks can decay individually and the B_c meson has a sufficiently large mass, one can study a great variety of decay channels. There have been many theoretical efforts to calculate the semileptonic and non-leptonic decays of the B_c meson. The semileptonic B_c decays provide a good opportunity to measure not only the Cabibbo-Kobayashi-Maskawa(CKM) elements but also the

* Corresponding author.

M.-W. Han and J. Lee (Eds.): EKC 2010, SPPHY 138, pp. 59–66.
springerlink.com

weak form factors for the transitions of B_c to bottom and charmed mesons. The nonleptonic B_c decays, in which only hadrons appear in the final state, are strongly influenced by the confining color forces among the quarks. In our recent paper [3,4], we analyzed the semileptonic[3] and rare[4] B_c decays using our light-front quark model(LFQM) based on the QCD-motivated effective LF Hamiltonian [5]. The weak form factors for the semileptonic and rare decays between two pseudoscalar mesons are obtained in the spacelike region and then analytically continued to the physical timelike region. The covariance (i.e. frame independence) of our model has been checked by performing the LF calculation in parallel with the manifestly covariant calculation using the exactly solvable covariant fermion field theory model in (3+1) dimensions.

In this paper, we extend our previous LFQM analysis of the semileptonic B_c decays to the nonleptonic two-body decays of B_c mesons. The QCD factorization approach is widely used since it works reasonably well in heavy-quark physics [6]. The factorization approximates the complicated non-leptonic decay amplitude into the product of the meson decay constant and the form factor. We shall use the form factors for semileptonic decays as well as the meson decay constants obtained in our LFQM as input parameters for the nonleptonic B_c decays.

The paper is organized as follows. In Sec. 2, we discuss the weak Hamiltonian responsible for the nonleptonic two-body decays of the B_c meson. In Sec. 3, we briefly describe our LFQM. Section 4 is devoted to the numerical results. A summary and conclusions follow in Sec.5.

2 Nonleptonic Two-Body B_c Decays

The nonleptonic weak decays are described in the standard model by a single W boson exchange diagram at tree level. In the standard model, the nonleptonic B_c decays are described by the effective Hamiltonian, which was obtained by integrating out the heavy W boson and top quark. For the case of $b\to c,u$ transition at the quark level, neglecting QCD penguin operators, one gets the following effective weak Hamiltonian:

$$H_{eff}^{b\to c(u)} = \frac{G_F}{\sqrt{2}}\{V_{cb}[c_1(\mu)O_1^{cb} + c_2(\mu)O_2^{cb}] + V_{ub}[c_1(\mu)O_1^{ub} + c_2(\mu)O_2^{ub}] + H.c\}, \tag{1}$$

where G_F is the Fermi coupling constant and $V_{q_1q_2}$ are the corresponding CKM matrix element. We use the central values of the CKM matrix elements quoted by the Particle Data Group (PDG)[7]. The effective weak Hamiltonian for the $c\to s,d$ transition can be defined in the same way. The effective weak Hamiltonian consists of products of local four-quark operators $O_{1(2)}$ renormalized at the scale μ, and scale-dependent Wilson coefficients $c_{1(2)}(\mu)$, which incorporate the short-distance effects arising from the renormalization of the effective Hamiltonian.

Without strong-interaction effects, one would have $c_1 = 1$ and $c_2 = 0$. However, this simple result is modified by gluon exchange: i.e., the original weak vertices get renormalized and the new types of interactions(such as the operators O_2 are induced. In these decays, the final hadrons are produced in the form of pointlike color-singlet objects with a large relative momentum. Thus, the hadronization of the decay products occurs after they separate far away from each other. This provides the possibility to avoid the final state interaction.

In the factorization approach to nonleptonic meson decays, one can distinguish three classes of decays [6]: (1) Class I: a color-favored transition and contains those decays in which only a charged meson can be generated directly from a color-singlet current, (2) Class II: a color-suppressed transition and contains those decays in which only a neutral meson can be generated directly from a color-singlet current, and (3) Class III: a transitions consists of those decays in which both Class I and II amplitudes interfere.

In this paper, we consider the following type of nonleptonic $B_c \to F_1 + F_2$ where F_1 is the pseudoscalar meson and F_2 the meson(vector or pseudoscalar) being factored out. The matrix elements of the semileptonic $B_c \to F_1$ decays can be parameterized by two Lorentz-invariant form factors:

$$\langle F_1(P_2) | V^\mu | B_c(P_1) \rangle = f_+(q^2) P^\mu + f_-(q^2) q^\mu, \tag{2}$$

where $P^\mu = P_1{}^\mu + P_2{}^\mu$ and $q^\mu = P_1{}^\mu - P_2{}^\mu$ is the four-momentum transfer to the lepton pair. Sometimes it is useful to express the matrix element of the vector current in terms of $f_+(q^2)$ and $f_0(q^2)$, which correspond to the transition amplitudes with 1^- and 0^+ spin-parity quantum numbers in the center of mass of the lepton pair, respectively. They satisfy the following relation:

$$f_0(q^2) = f_+(q^2) + \frac{q^2}{M_1^2 - M_2^2} f_-(q^2).\cdot \tag{3}$$

The decay constants of pseudoscalar and vector mesons are defined by

$$\begin{aligned} \langle P(p) | A^\mu | 0 \rangle &= -i f_P p^\mu, \\ \langle V(p,h) | V^\mu | 0 \rangle &= f_V M_V \varepsilon^\mu(h), \end{aligned} \tag{4}$$

where $\varepsilon(h)$ is the polarization vector of the vector meson.

Using Eqs. (2) and (4) and the factorziation approximation, the nonleptonic decay amplitudes can be expressed as the product of one particle matrix elements. For example, the nonleptonic decay amplitude for the class I $b \to (u,c)(q_1 \bar{q}_2)$ process is obtained by

$$\langle \eta_c M^+ | H_{eff} | B_c^+ \rangle = \frac{G_F}{\sqrt{2}} V_{cb} V^*_{q_1 q_2} a_1 X^{(B_c^+ \eta_c, M^+)}, \tag{5}$$

where

$$X^{(B_c^+ \eta_c, M^+)} = -i f_{\eta_c} (M_{B_c}^2 - M_{\eta_c}^2) f_0^{B_c - \eta_c}(M_{M^+}^2), \tag{6}$$

and a_1 is the Wilson coefficient. Other classes of decay amplitudes can be found in [8].

The decay rate for $B_c \to F_1 + F_2$ in the rest frame of the B_c meson is given by

$$\Gamma(B_c \to F_1 F_2) = \frac{p_c}{8\pi M_{B_c}^2} |\langle F_1 F_2 | H_{eff} | B_c \rangle|^2, \tag{7}$$

where

$$p_c = \frac{\sqrt{[M_{B_c}^2 - (M_1 + M_2)^2][M_{B_c}^2 - (M_1 - M_2)^2]}}{2M_{B_c}} \tag{8}$$

is the center of mass momentum of the final state meson.

3 Model Description

The key idea in our LFQM [5] for mesons is to treat the radial wave function as a trial function for variational principle to the QCD-motivated effective Hamiltonian saturating the Fock state expansion by the constituent quark and antiquark. The QCD-motivated Hamiltonian for a description of the ground state meson mass spectra is given by

$$\begin{aligned} H_{q\bar{q}} |\Psi\rangle &= \left[\sqrt{m_q^2 + \vec{k}^2} + \sqrt{m_{\bar{q}}^2 + \vec{k}^2} + V_{q\bar{q}} \right] |\Psi\rangle \\ &= [H_0 + V_{q\bar{q}}] |\Psi\rangle = M |\Psi\rangle, \end{aligned} \tag{9}$$

where $\vec{k} = (k_\perp, k_z)$ is the three-momentum of the constituent quark, M is the mass of the meson, and Ψ is the meson wave function. In this work, we use the interaction potential $V_{q\bar{q}}$ as Coulomb plus linear confining potentials. The hyperfine interaction essential to distinguish pseudoscalar and vector mesons is also included; viz.,

$$V_{q\bar{q}} = V_0 + V_{hyp} = a + br - \frac{4\alpha_s}{3r} + V_{hyp}. \tag{10}$$

The momentum-space light-front wave function of the ground state pseudoscalar and vector mesons is given by

$$\Psi(x_i, k_{i\perp}, \lambda_i) = R_{\lambda_1 \lambda_2}(x_i, k_{\perp i}) \phi(x_i, k_{i\perp}), \tag{11}$$

where ϕ is the radial wave function and R is the spin orbit wave function. The model wave function in Eq. (11) is represented by the Lorentz-invariant internal variables, $x_i = p^+_i / P^+$, $k_{i\perp} = p_{i\perp} - x_i P_\perp$, where $P = (P^+, P^-, P_\perp)$ is the momentum of the meson M, and p_i and λ_i are the momenta and the helicities of constituent quarks, respectively. For the radial wave function, we use $\phi \propto \exp(-\vec{k}^2/2\beta^2)$ for both pseudoscalar and vector mesons. We apply our variational principle to the QCD-motivated effective Hamiltonian first to evaluate the expectation value of the central Hamiltonian H_0+V_0 with a trial function that depends on the variational parameter. Once the model parameters are fixed by minimizing the expectation value of H_0+V_0, then the mass eigenvalue of each meson is obtained as $M = \langle \phi | H_0 + V_{q\bar{q}} | \phi \rangle$. A more detailed procedure for determining the model parameters of light- and heavy-quark sectors can be found in our previous works [5].

In this work, we shall also consider the nonleptonic decays of B_c mesons to isoscalar states such as ω and (η, η'). Isoscalar states with the same J^{PC} will mix, but mixing between the two light-quark isoscalar mesons, and the much heavier charmonium or bottomonium states is generally assumed to be negligible. Since the vector mixing angle is known to be very close to ideal mixing, we assume ideal mixing between ω and ϕ mesons. However, the octet-singlet mixing angle θ of (η, η') is known to be in the range of -10° to -23°. The physical η and η' are the mixtures of the flavor SU(3) $\eta_8 = (u\bar{u} + d\bar{d} - 2s\bar{s})/\sqrt{6}$ and $\eta_0 = (u\bar{u} + d\bar{d} + s\bar{s})/\sqrt{3}$ states:

$$\begin{pmatrix} \eta \\ \eta' \end{pmatrix} = \begin{pmatrix} \cos\theta & -\sin\theta \\ \sin\theta & \cos\theta \end{pmatrix} \begin{pmatrix} \eta_8 \\ \eta_0 \end{pmatrix}. \tag{12}$$

Analogously, in terms of the quark-flavor basis $\eta_q = (u\bar{u} + d\bar{d})/\sqrt{2}$ and $\eta_s = s\bar{s}$, one obtains[9]

$$\begin{pmatrix} \eta \\ \eta' \end{pmatrix} = \begin{pmatrix} \cos\phi & -\sin\phi \\ \sin\phi & \cos\phi \end{pmatrix} \begin{pmatrix} \eta_q \\ \eta_s \end{pmatrix}. \tag{13}$$

The two schemes are equivalent to each other by $\phi = \theta + \arctan\sqrt{2}$ when SU(3) symmetry is perfect. In our previous work[5], we obtained the $\eta - \eta'$ mixing angle $\theta =$ -19° by fitting the physical masses of η and η'. We applied this mixing angle to obtain the decay constants related to η and η' mesons as follows[8]: $f_\eta^q = 103.2$ MeV, $f_\eta^s = -116.6$ MeV, $f_{\eta'}^q = 74.2$ MeV, and $f_{\eta'}^s = 155.3$ MeV.

4 Numerical Results

In our numerical calculations of exclusive B_c decays, we use the model parameters (m, β) for the linear confining potentials given in [3,4,5] to compute the weak form factors for semileptonic B_c decays. In Table 1, we show the decay form factors $f_+(0)=f_0(0)$ at $q^2=0$ for the semileptonic $B_c \to (D_{(s)}, \eta_c, B_{(s)})$ decays obtained from [3,4].

Table 1 Form factors $f_+(0)=f_0(0)$ at q_2=0 for the semileptonic B_c decays

Mode	$f_+(0)=f_0(0)$
$B_c \to D$	0.086
$B_c \to D_s$	0.120
$B_c \to \eta_c$	0.482
$B_c \to B$	0.467
$B_c \to B_s^0$	0.573

Using them together with the CKM matrix elements, we predict the branching ratios which are given in Table 2. The analysis of the more comprehensive decay modes can be found in [8].

Table 2 Branching ratios(in %) of the exclusive nonleptonic B_c decays

Class	Mode	Br (%)
I	$B_c^+ \to \eta_c \pi^+$	0.091
	$B_c^+ \to \eta_c \rho^+$	0.257
	$B_c^+ \to B^0 \pi^+$	0.157
	$B_c^+ \to B^0 \rho^+$	0.195
	$B_c^+ \to B_s^0 \pi^+$	3.723
	$B_c^+ \to B_s^0 \rho^+$	2.561
II	$B_c^+ \to B^+ \pi^0$	5.5 x 10^{-3}
	$B_c^+ \to B^+ \rho^0$	6.8 x 10^{-3}
	$B_c^+ \to B^+ \eta$	2.8 x 10^{-2}
	$B_c^+ \to B^+ \eta'$	3.8 x 10^{-4}
	$B_c^+ \to B^+ \overline{\mathrm{K}}^0$	0.336
	$B_c^+ \to B^+ \mathrm{K}^0$	8.8 x 10^{-4}
III	$B_c^+ \to D^+ D^0$	7.5 x 10^{-6}
	$B_c^+ \to D_s^+ D^0$	2.0 x 10^{-4}
	$B_c^+ \to D^+ \eta$	0.014
	$B_c^+ \to D_s^+ \eta$	0.378

The relative size of the branching ratios for various decay modes may be estimated from power counting of the Wilson coefficients and the CKM factors with respect to the small parameter of the Cabibbo angle $\lambda = \sin\theta_C$ in the Wolfenstein parametrization[10], e.g. the CKM matrix elements can be expanded in terms of λ as $V_{ud} \sim 1, V_{cs} \sim \lambda$, $V_{ub} \sim \lambda^3$, $V_{cd} \sim -\lambda$, $V_{cs} \sim 1$, and $V_{cb} \sim \lambda^2$. From Table 1, we make the following observations:

(1) The class I decay modes have comparatively large branching ratios. The CKM favored c decays such as $B_c^+ \to B_s^0(\pi^+, \rho^+)$ decays with the CKM factor $V_{cs}V_{ud}^* \sim \lambda^0$ have branching ratios of the order of 10^{-2}, which are the most promising class I decay modes shown in Table 2.

(2) The branching ratios for the class II decay modes are relatively smaller than those for the class I decay modes. However, the $B_c^+ \to B^+\overline{K}^0$ decay with $V_{cs}V_{ud}^* \sim \lambda^0$ has branching ratio of the order of 10^{-3} and should be accessible experimentally. Of interest is the abnormally small branching ratio of $B_c^+ \to B^+\eta'$ compared to that of $B_c^+ \to B^+\eta$. The reason for such a small branching ratio is not only because the available physical phase space is too small but also because there are large destructive interferences between η'_q and η'_s due to the serious cancellation between the CKM factors $V_{cd}V_{ud}^*$ and $V_{cs}V_{us}^*$.

(3) The class III decay modes involve the Pauli interference. Especially, the interference is the most significantly involved in the $B_c^+ \to D_s^+D^0$ decay compared to others.

5 Summary

In this work, we have studied the exclusive nonleptonic $B_c \to (D_{(s)}, \eta_c, B_{(s)})M$ decays where the final state M mesons are factored out in the QCD factorization approach. The inputs used to obtain their branching ratios were the weak form factors for the semileptonic $B_c \to (D_{(s)}, \eta_c, B_{(s)})$ decays in the whole kinematical region and the unmeasured weak decay constants obtained from our previous LFQM analysis [3,4,5,8]. For the measured values of decay constants, we use the central values extracted from the experimental measurements[7]. Our predictions for the branching ratios is summarized in Table 2. The upcoming experimental measurements of the corresponding decay rates can examine our predictions. The most promising measurable decay modes appear to be the CKM favored c decays such as $B_c^+ \to B_s^0(\pi^+, \rho^+)$ decays. It is thus expected that the dominant contribution to the B_c total rate comes from the c induced decays. The more c induced $B_c \to VP$ and $B_c \to VV$ decay modes seem to deserve further consideration.

References

[1] Abe, F., et al.: CDF Collaboration. Phys.Rev. Lett. 81, 24–32 (1998)
[2] Gouz, I.P., et al.: Yad. Fiz. 67, 1581 (2004)
[3] Choi, H.-M., Ji, C.-R.: Phys. Rev. D 80, 054016 (2009)
[4] Choi, H.-M.: Phys. Rev. D 81, 054003 (2010)
[5] Choi, H.-M., Ji, C.-R.: Phys. Rev. D 59, 074015 (1999) Phys. Lett. B 460, 461 (1999)
[6] Bauer, M., Stech, B., Wirbel, M.: Z. Phys. C 34, 103 (1987)
[7] Amsler, C., et al.: Particle Data Group. Phys. Lett. B 667 (1008) 1
[8] Choi, H.-M., Ji, C.-R.: Phys. Rev. D 80, 114003 (2009)
[9] Feldmann, T., Kroll, P., Stech, B.: Phys. Rev. D 58, 114006 (1998)
[10] Wolfenstein, L.: Phys. Rev. Lett. 51, 1945 (1983)

Design of a Totally Implantable Artificial Cochlea Mimicking the Human Hearing Mechanism

W.D. Kim*, J.H. Lee, H.S. Choi, S. Hur, and J.S. Park

Korea Institute of Machinery & Materials
wdkim@kimm.re.kr

Abstract. In this research, a bio-inspired artificial cochlea will be designed based on the artificial basilar membrane and the nanowire mimicking stereocilia. It has the potential to be used for the totally implantable artificial cochlea.

We developed the FEM models to simulate the behavior of the artificial basilar membrane to understand the frequency separation mechanism of the human basilar membrane. The exponentially tapered membranes with several thicknesses were designed using silicon nitride (Si_3N_4) and polyimide (PI). All the exponentially tapered membranes successfully showed the frequency shifting effect and the models were effectively used for optimizing the design of the artificial basilar membrane.

We fabricated the ZnO nanowires mimicking the stereocilia in various sizes and measured the piezoelectric potential output of the ZnO nanowires. The mechanism of the ZnO nanowires was also understood through the FEM simulation. It was concluded that ZnO nanowires can generate electric current by the piezoelectric effect and ZnO nanowires have potential to be used as a power source for nano-scale devices.

To conclude, the work is concentrated on the design of the artificial basilar membrane and the piezoelectric ZnO nanowires for the artificial mechanosensory system.

1 Introduction

The human ear consists of the outer ear, the middle ear, and the inner ear. The external auditory canal, a tube running from the outer ear to the middle ear, directs sound waves from the auricle to the tympanic membrane (ear drum). The three small bones (malleus, incus, and stapes) in the middle ear transmit the sound waves to the membranous oval window, the intersection between the middle ear and the inner ear. The sound waves travel to the brain along the auditory nerve in the cochlea of the inner ear. The cochlea has two core mechanosensory components: the basilar membrane and the hair cells having the stereocilia. The basilar membrane separates the incoming sound waves by their frequencies [1]. The hair cells are on the basilar membrane and act as the sensory receptors generating bioelectric signals [1].

* Corresponding author.

M.-W. Han and J. Lee (Eds.): EKC 2010, SPPHY 138, pp. 67–75.
springerlink.com

At present, the artificial cochlear implementation is the most promising solution for the hearing loss caused by the cochlear damage. The conventional artificial cochlea has the internal and external components; the external parts consist of a battery, microphone, speech processor, coil and the internal parts consist of receiver, electrodes to stimulate the hearing nerves. Therefore, the existing cochlear implant has several major disadvantages such as the disclosure of disability, large power consumption requiring frequent recharge, and the inconvenience of carrying some external components.

In this research, a bio-inspired artificial cochlea will be designed based on the artificial basilar membrane (ABM) and the stereocilia. It has the potential to be used for the totally implantable artificial cochlea that requires minimized or no external power as shown in Fig. 1.

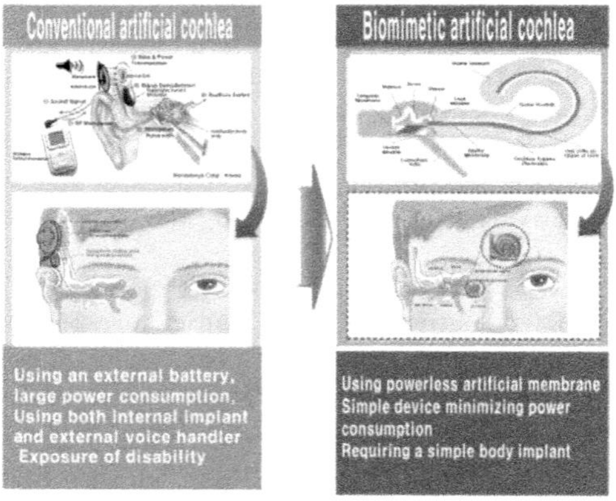

Fig. 1 Comparison of biomimetic artificial cochlea with conventional artificial cochlea

2 Abm Modeling

We developed the FEM models to simulate the behavior of biomimetic, new-concept ABM to understand the frequency separation mechanism of the human basilar membrane. Exponentially tapered membranes with 1.4μm and 4 μm thickness were designed. The materials used for the models are silicon nitride (Si_3N_4) and polyimide (PI). Each model has a small rectangular membrane for the acoustic input of one Pascal. For the simulation, five different input frequencies are used: 4.2, 7.2, 12.3, 20, and 35 all in kHz [2].

The residual stress in the membrane was expressed as the tension applied to the membrane for the understanding of the residual stress effects on the frequency separation. The following two cases were used for tension; 1) No tension and 2) T_x=15N/m, T_y=120N/m. FEM models to model an ABM for the understanding of the behavior of the human basilar membrane is developed using COMSOL^{TM}. The schematic of the model presented in Fig. 2.

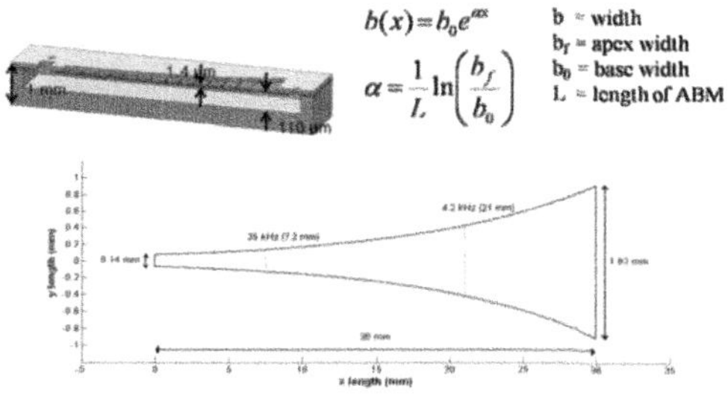

Fig. 2 The schematic of the artificial basilar membrane model

The conditions for the cases simulated using COMSOL™ are shown on Table 1.

Table 1 The conditions for the simulation

Material	Thickness	Case 1	Case 2
Si_3N_4	1.4 µm	No residual stress (no tension)	T_x=15 N/m (10.7 MPa) T_y=120 N/m (85.7 MPa)
	4.0 µm		T_x=15 N/m (3.75 MPa) T_y=120 N/m (30 MPa)
Polyimide	1.4 µm		T_x=15 N/m (10.7 MPa) T_y=120 N/m (85.7 MPa)
	4.0 µm		T_x=15 N/m (3.75 MPa) T_y=120 N/m (30 MPa)

The displacement responses of ABM models as the function of applied frequency are presented in Fig. 3 for 1.4 µm, Si_3N_4 membrane and the 3D view of the membrane is shown in Fig. 4.

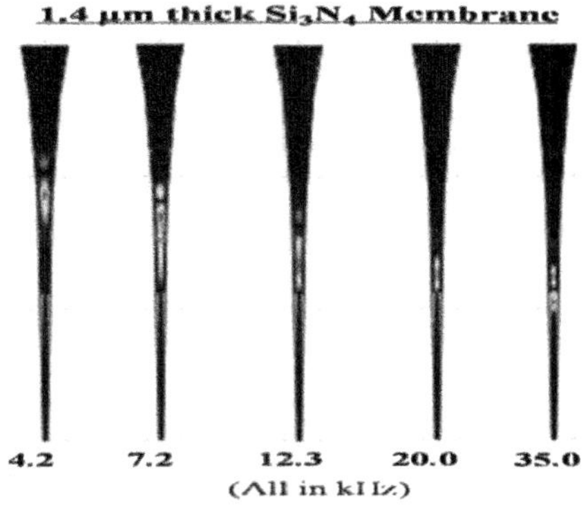

Fig. 3 The displacement responses of the 1.4 µm, Si_3N_4 artificial basilar membrane model as a function of input frequencies

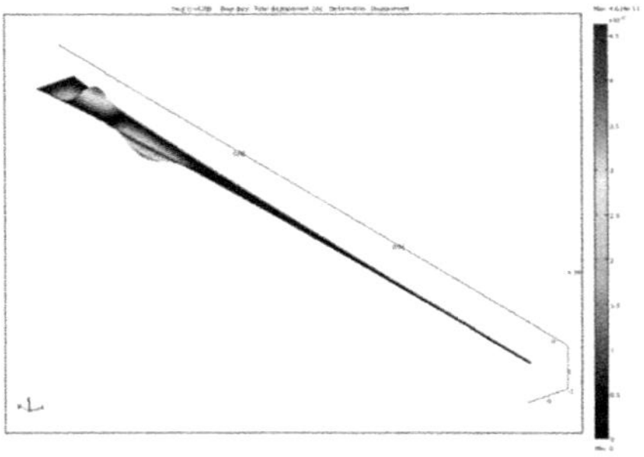

Fig. 4 The 3D view of membrane displacement

The displacement responses of 1.4 μm and 4 μm thick Si_3N_4 membranes as functions of symmetric axis distance are shown on Figs. 5 and 6.

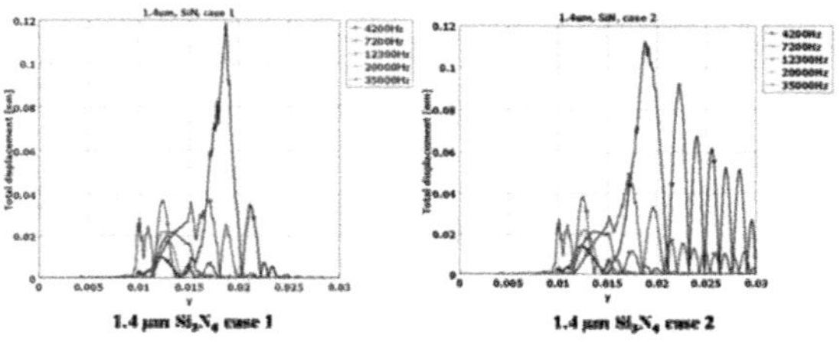

Fig. 5 The displacement responses of the 1.4 μm, Si_3N_4 artificial basilar membrane model as a function of symmetric axis distance

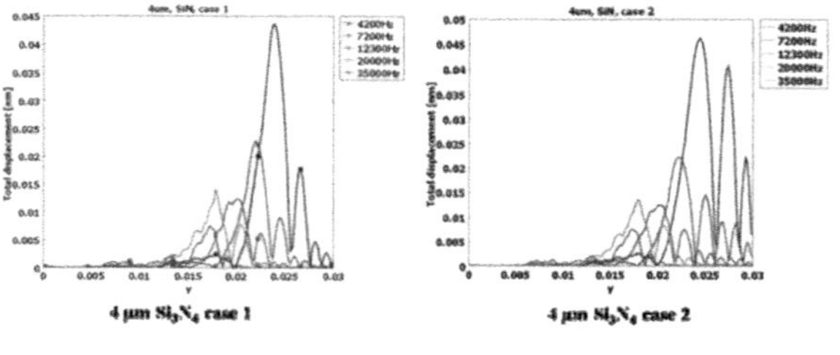

Fig. 6 The displacement responses of the 4 μm, Si_3N_4 ABM model as a function of symmetric axis distance.

As shown on above figures, the 4 μm, Si_3N_4 membrane has less effect on the residual stress. The frequencies of Si_3N_4 membranes are not significantly shifted by the residual stress due to the high stiffness of Si_3N_4. Due to its high stiffness, the vibration at the point of the maximum displacement is not dispersed to adjacent area, but the value of the total displacement is small.

Figs. 5 and 6 display the displacement responses of 1.4 μm and 4 μm thick Si_3N_4 membranes as functions of symmetric axis distance. The displacement for PI membrane also presented in Figs. 7 and 8 for 1.4 μm and 4 μm thicknesses. PI membranes show better frequency shifting effect as the function of the applied frequency due to its lower stiffness.

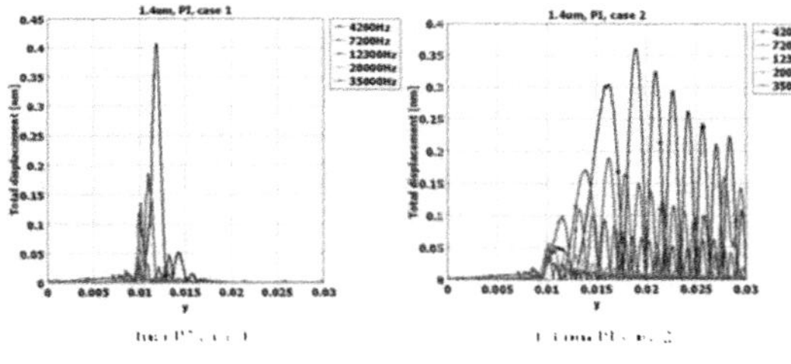

Fig. 7 The displacement responses of the 1.4 μm, PI ABM model as a function of symmetric axis distance

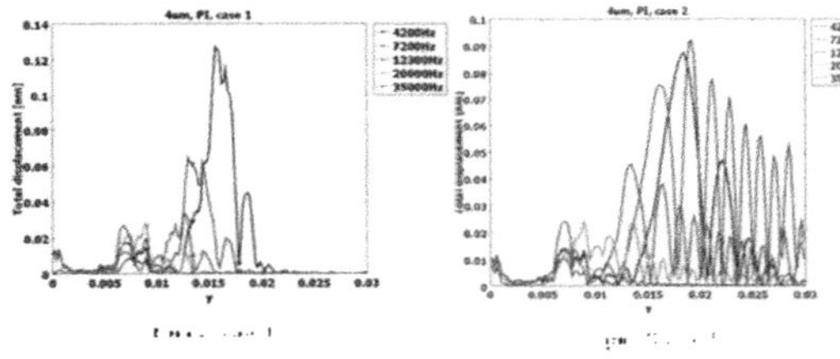

Fig. 8 The displacement responses of the 1.4 μm, PI ABM model as a function of symmetric axis distance

The frequencies of PI membranes are noticeably shifted by the residual stress compared to those of Si_3N_4 membranes. Due to its low stiffness, the value of the total displacement is great, but the vibration at the point of the maximum displacement is easily dispersed to adjacent area that frequency separation is not accomplished.

From ABM modeling, it was concluded that the stiffness of material and thickness of membrane should be optimized to maximize frequency separation and increase the displacement of membrane.

3 Modeling and Experimental Procedures of ZnO Nanowire

ZnO nanowires are studied to mimic the operating mechanism of ion channels by stereocilia. ZnO nanowires are typical piezoelectric materials used for electromechanical and thermo-electrical energy conversion. We fabricated the ZnO nanowires mimicking the stereocilia in various sizes and measured the piezoelectric potential output of the ZnO nanowires when they were deflected by a Current Atomic Force Microscope (I-AFM) tip in contact mode. The mechanism of the ZnO nanowires were also understood through the FEM simulation and the simulated results were compared with the experimental results.

To understand how the ZnO nanowires generate an output signal while bending, FEM simulation was performed using COMSOL™ Multiphysics [3-5]. The ZnO nanowires were modeled as perfect cylinders and two different sizes were used for the simulation. The dimensions of the two nanowires are as follows: 1) D=50nm, L=600nm and 2) D=300nm, L=2000nm. The magnitudes of the applied lateral forces were varied for each problem. The calculated result for the nanowire of D=50nm, L=600nm with the applied force of 80nN are presented in Fig. 9. Red color indicates positive electric potential and blue color denotes negative electric potential. Fig. 9(b) shows the distribution of electric potential on the top surface (Z=600nm) of the ZnO nanowire ranging from -0.3V to +0.3V.

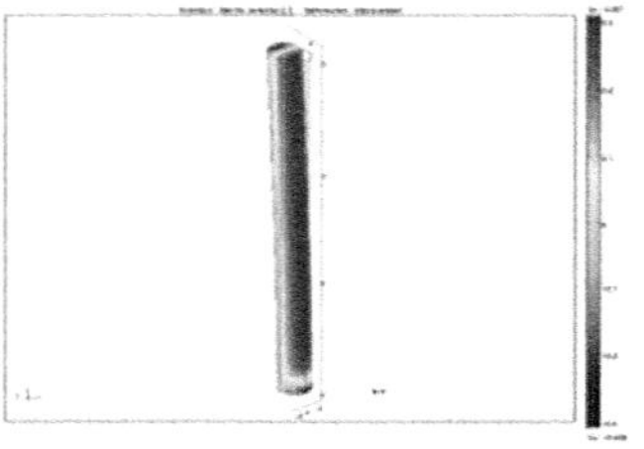

(a)

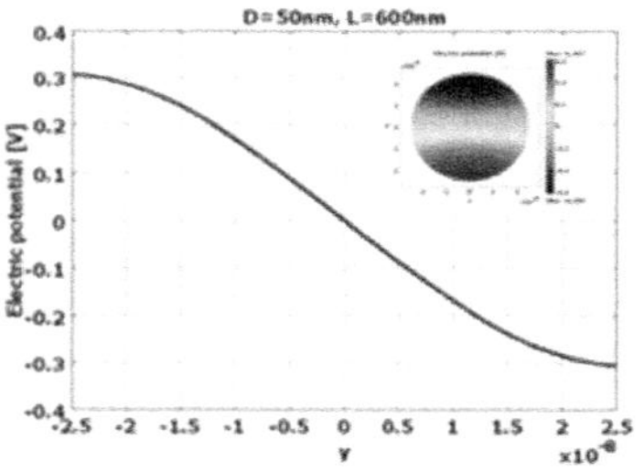

(b)

Fig. 9 (a) The electric potential distribution of the ZnO nanowire and (b) The electric potential on the top surface, Z=600nm (D = 50nm, L = 600nm, Applied force: Fv = 80nN)

Fig. 10 shows the electric potential at the side of the ZnO nanowire generating the minimum electric potential. It is important to note that the potential distribution on the ZnO nanowire is shifted near the bottom of the nanowires.

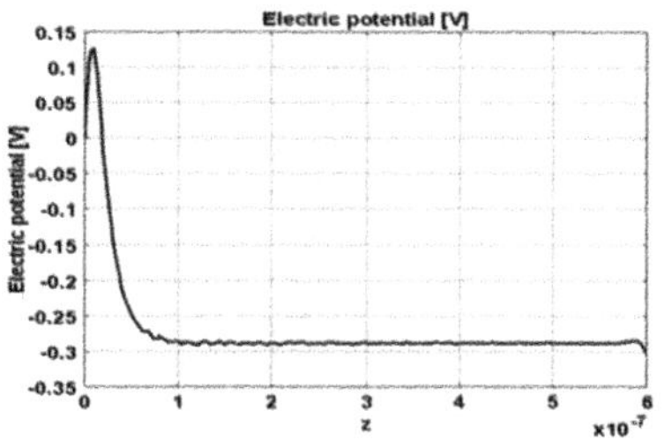

Fig. 10 Shifted electric potential near the bottom of the nanowire

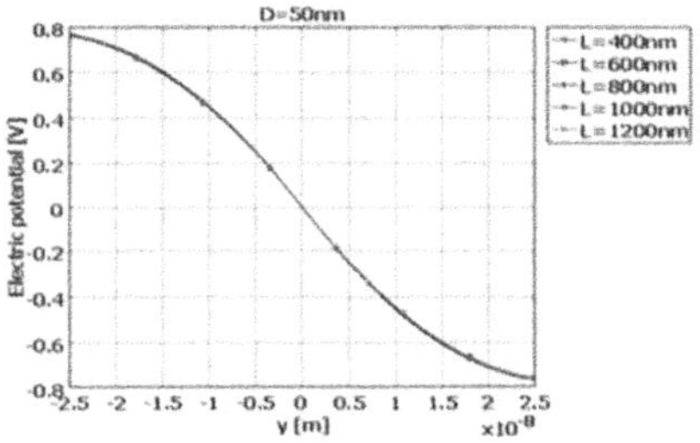

Fig. 11 The electric potentials at the cross sections of the ZnO nanowires with different lengths (D = 50nm, L = 400~1200nm, Applied force: Fv = 200nN)

The electric potential distribution produced on the ZnO nanowire of D= 50nm with length varying from 400 to 1200nm was simulated. Fig. 11 displays the values of the electric potential at the cross section of the nanowires ranging from -0.8V to +0.8V with the applied force of 200nN. The figure also shows that the generated electric potential was not changed regardless of varying lengths of the ZnO nanowire. Therefore, it can be concluded that the electric potential is affected more by the applied force than the length of the ZnO nanowire.

The ZnO nanowires have been fabricated on the Si substrate by the solution method. The ZnO thin film was used as the seed layer to grow ZnO nanowires. The ZnO thin film was prepared on the Si substrate by RF magnetron sputtering. The ZnO disc with 99.99% purity was used as the target. The mixtures of 99.999% pure Ar and O_2 gases were used as the reaction gases. The Ar and O_2 flow rates were 16 and 64 sccm (Standard Cubic Centimeter per Minute), respectively. The substrate temperature was kept at room temperature during deposition. The RF sputtering power was set to 240W, and the sputtering pressure was 20mTorr. The sputtering process time was 33min. to get ~60nm ZnO thin film. The ZnO nanowires were synthesized in an aqueous solution at 80℃. In this procedure, aqueous solutions of zinc nitrate hexahydrate ($Zn(NO_3)_2.6H_2O$) and hexa-methylenetetramine ($(CH_2)_6N_4$) were mixed with DI water and refluxed for 6 hours [6]. We used an I-AFM to measure the generated current at the ends of ZnO nanowires while pushing the nanowires using AFM tip. The generated current is shown on Figs. 12 and 13 in 3D and 2D and there are current peaks at each ZnO nanowire. We noticed that the maximum generated current was around 90pA.

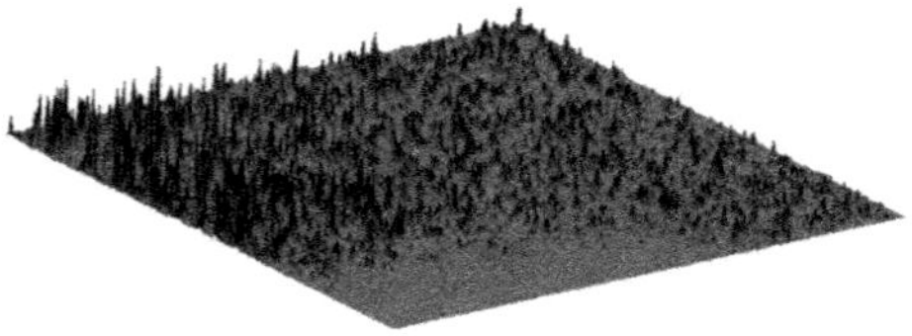

Fig. 12 Current of a ZnO nanowire using I-AFM (3D)

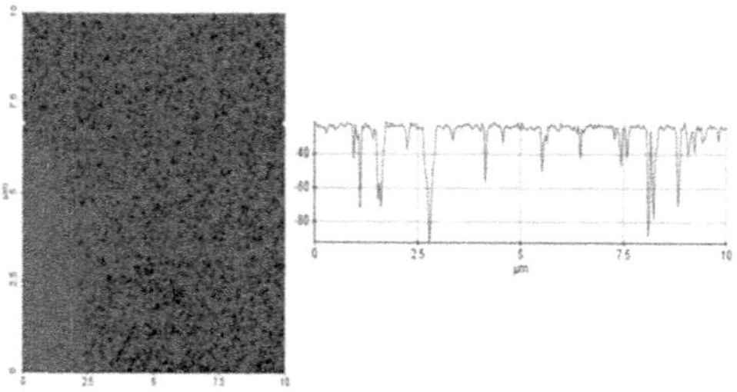

Fig. 13 Current of a ZnO nanowire using I-AFM (2D)

Based on this experiment, we can conclude that ZnO nanowires can generate electric current by the piezoelectric effect. It shows the potential to use ZnO nanowires as a power source for nano-scale devices.

4 Conclusions

We successfully developed FEM models to understand the behavior of our ABM and showed the possibility to optimize the design of our ABM using the developed FEM models. Also, it was concluded that ZnO nanowires can generate electric current by the piezoelectric effect through the simulation and measurement using I-AFM. It shows the potential to use ZnO nanowires as a power source for nano-scale devices such as the bio-inspired artificial cochlea.

To conclude, the work is concentrated on the design of the artificial basilar membrane and the piezoelectric ZnO nanowires for the artificial mechanosensory system.

Acknowledgement. This research was supported by the Pioneer Research Center Program through the National Research Foundation of Korea funded by the Ministry of Education, Science and Technology (2010-0018347/0002248).

References

[1] Robles, L., Ruggero, M.A.: Mechanics of the Mammalian Cochlea. Physiological Reviews 81(3), 1305–1352 (2001)

[2] White, R., Grosh, K.: Microengineered hydromechanical cochlear model. In: Proceedings of the National Academy of Sciences, vol. 102(5), pp. 1296–1301 (February 2005)

[3] Gao, Y., Wang, Z.L.: Equilibrium potential of free charge carriers in a bent piezoelectric semiconductive nanowire. Nano Letters 9(3), 1103–1110 (2009)

[4] Gao, Y., Wang, Z.L.: Electrostatic potential in a bent piezoelectric nanowire. The fundamental theory of nanogenerator and nanopiezotronics. Nano Letters 7(8), 2499–2505 (2007)

[5] Wang, Z.L.: Towards self-powered nanosystems: From nanogenerators to nanopiezotronics. Adv. Funct. Mater. 18, 3553–3567 (2008)

[6] Reddy, N.K., Ahsanulhaq, Q., Kim, J.H., Hahn, Y.B.: Well-aligned ZnO nanorods for device applications: Synthesis and characterization of ZnO nanorods and n-ZnO/p-Si heterojunction diodes. Europhysics Letters 81, 38001 (2008)

A Study on Young's Internet Overindulgence Testing Method with Lisrel and Clustering Algorithms

Chan Jung Park[1] and Jung Suk Hyun[2,*]

[1] Dept. of Computer Education, Jeju National University, Jeju-do, Republic of Korea
[2] Dept. of Management Information Systems, Jeju National University, Jeju-do, Republic of Korea
jshyun@jejunu.ac.kr

Abstract. Due to the advance of the Internet technologies, the Internet addition of adolescents has become a hot issue in our society. Thus, Cyber ethics education in secondary schools has become important. One of the issues covered in Cyber Ethics education is the Internet overindulgence. To find out the main reason for the Internet overindulgence is a key issue in Cyber ethics education. Many organizations use Young's questionnaire when they measure their students' overindulgence levels. In addition, they guide the students referring to their overindulgence levels. However, several research works began to point out some problems of the Young's measurement in the aspect of statistics. In this paper, we analyze the problems of Young's research on the Internet overindulgence testing methods and propose a new way to test by using the techniques such as Lisrel and clustering algorithms. In order to accomplish these, we survey on 334 secondary school students with the Young's questionnaire and analyze results with Weka tool.

1 Introduction

As the Internet is more common, adolescent's Internet overindulgence has become a serious problem recently. There have been a large number of research works about the Internet overindulgence so far. Among them, Dr. Young's research and his testing items [1][2] are the most commonly used when people diagnose their levels of addiction [3][4].

Recently, a few research works found out some problems in Young's method. According to Kang et al.'s research [5], even though there exists several more important testing items than others in Young's testing items, Young regarded 20 testing items equally. Kang et al. proved Young's problem with the factor analysis. According to Kang et al., the number of the lower factors included in the reduced factors are not the same. It means that each factor does not have the same importance. In addition, there exists correlation between Young's testing items. Thus, to assign the same point to the problems is not fair and it causes double counting bias.

* Corresponding author.

M.-W. Han and J. Lee (Eds.): EKC 2010, SPPHY 138, pp. 77–83.
springerlink.com

In this paper, we analyze the problems of Young's research on the Internet overindulgence testing methods and propose a new way to test. In order to accomplish these, we survey on 334 secondary school students with the Young's questionnaire and analyze results with Lisrel and Weka tool. Lisrel is an acronym for linear structural relations. It is a statistical software package used in structural equation modeling [6]. On the other hand, Weka contains a collection of visualization tools and algorithms for data analysis and predictive modeling, together with graphical user interfaces for easy access to this functionality [7][8].

The rest of this paper is organized as follows. In Section 2, we introduce Young's testing items. In Section 3, we describe our experimental environment and the results of statistical analysis with factor analysis, Lisrel, and clustering algorithms. Finally, in Section 4, we conclude our paper.

2 Background

In this section, we describe Young's Internet overindulgence testing method for our experiment. The testing questions are as following:

c01: How often do you find that you stay on-line longer than you intended?
c02: How often do you neglect household chores to spend more time on-line?
c03: How often do you prefer the excitement of the Internet to intimacy with your partner?
c04: How often do you form new relationships with fellow on-line users?
c05: How often do others in your life complain to you about the amount of time you spend on-line?
c06: How often do your grades or school work suffer because of the amount of time you spend on-line?
c07: How often do you check your e-mail before something else that you need to do?
c08: How often does your job performance or productivity suffer because of the Internet?
c09: How often do you become defensive or secretive when anyone asks you what you do on-line?
c10: How often do you block out disturbing thoughts about your life with soothing thoughts of the Internet?
c11: How often do you find yourself anticipating when you will go on-line again?
c12: How often do you fear that life without the Internet would be boring, empty, and joyless?
c13: How often do you snap, yell, or act annoyed if someone bothers you while you are online?
c14: How often do you lose sleep due to late-night log-ins?
c15: How often do you feel preoccupied with the Internet when off-line, or fantasize about being on-line?
c16. How often do you find yourself saying "just a few more minutes" when on-line?
c17: How often do you try to cut down the amount of time you spend on-line and fail?
c18: How often do you try to hide how long you've been on-line?
c19: How often do you choose to spend more time on-line over going out with others?
c20: How often do you feel depressed, moody, or nervous when you are off-line, which goes away once you are back on-line?

After the test, a respondent obtains a final score. When we calculate the score, 20 items have the same weight. In other words, each item can have 1(rarely) point - 5(always) points. And then, we can add the 20 scores to find out the total. If the total score lies between 20-49 points, then the person is an average user. If the total score lies in 50-79, the person is experiencing occasional problems on the Internet. If the total score is higher than 80, the person is highly addicted to the Internet.

3 Experiment

In this section, we firstly describe our respondents and then, we analysis Young's measurement with factor analysis. Next, in order to overcome the weakness of the factor analysis, we run Lisrel to confirm our experiment. Finally, in order to see the characteristics for each overindulgence level by using clustering algorithms.

3.1 Analysis Sample

In our experiment, 334 respondents were answered to our questionnaire. The summary is shown in Table 1.

Table 1 Respondents

		No	Percentage
gender	Female	185	55.4%
	Male	149	44.6%
class	Middle school	126	37.7%
	High school	208	62.3%

3.2 Factor Analysis

Next, in our experiment, we performed a factor analysis as shown in Table 2. We extracted 4 factors : (1) the factor that was related to the *insufficiency of adaptation in real life* (c11, c12, c13, c14, c15, c19, c20), (2) the factor that was related to the lack of self-control (c1, c2, c16, c17), (3) the factor that was related to obsession (c5, c7, c9, c10, c18), and (4) the factor that was related to the preference to virtual reality (c3, c4, c6, c8).

From our experiment, 4 factors did not include the same amount of lower factors, and it means that among 4 factors, there exist more important factors than others.

Table 2 Factor analysis

	Factors			
	1	2	3	4
C15	.754	.005	.138	.024
C20	.716	.104	.065	.223
C19	.621	.272	.169	.177
C13	.604	.332	.140	.143
C12	.573	.209	.103	.307
C11	.489	.292	.337	.098
C14	.422	.318	.105	.315
C01	.145	.764	.102	.246
C02	.029	.710	.127	.307
C16	.406	.671	.212	-.035
C17	.384	.648	.202	-.051
C09	.194	.116	.735	.055
C18	.339	.241	.622	-.190
C05	-.020	.252	.562	.362
C10	.404	-.022	.544	.365
C07	.032	.052	.479	.080
C06	.115	.169	.316	.712
C08	.177	.154	.352	.695
C04	.203	.043	-.119	.509
C03	.334	.353	-.008	.410

3.3 Lisrel Analysis

In this Subsection, we analyze the problem about Young's measurement with Lisrel. Since Young's measurement has the assumption that each testing item is independent from each other, the scores about 20 items are added. And then, we use the total score when we classify overindulgence levels.

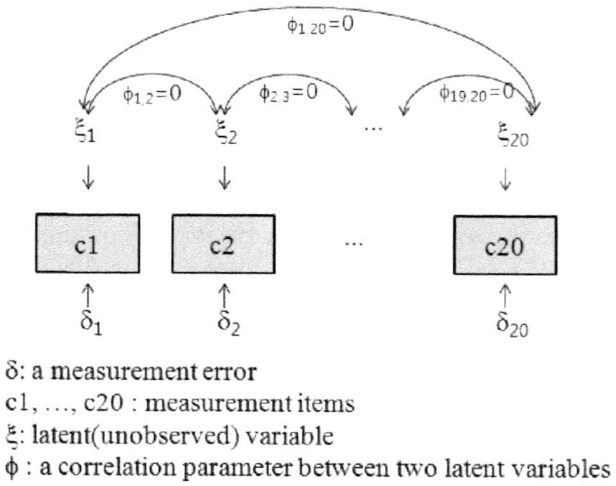

Fig. 1 A measurement items independence assumption model

Under this assumption, we used LISREL 8 with co-variances as input to estimate the model. Next, we performed a confirmatory factor analysis to assess whether a measurement items independence assumption model adequately accounted for the co-variances among the subset of items for each construct. A measurement items independence assumption model (see also Figure 1) had $\chi 2$ (190) = 2289.6; $p < .01$), the goodness-of-fit index (GFI) value was .358, and the adjusted goodness-of-fit index (AGFI) value was .290.

On the contrary, factor measurement items dependence assumption model had $\chi 2$ (168) = 1069.8; $p < .01$) the GFI = .664, AGFI = .580, respectively, which implies that the measurement items are not independent. In addition to this, chi-square values dramatically decreased from 2289.6 to 1069.8 owing to free correlation parameters between factors. The chi-square difference test between independence and dependence assumptions showed $\chi 2$ (22) = 1219.8; $p < .01$. The independence assumption between measurement items is rejected.

3.4 Clustering Analysis

Next, in order to find out the characteristics of each overindulgence level, we performed a clustering analysis with EM and K-means algorithms. The result of EM analysis is shown in Table 3.

Table 3 Analysis with EM clustering algorithm

Scheme: weka.clusterers.EM -I 100 -N -1 -M 1.0E-6 -S 100

Items	clusters				
	0 (0.19)	1 (0.23)	2 (0.29)	3 (0.22)	4 (0.07)
c1	3.67	3.59	2.90	4.13	**4.78**
c2	3.17	2.92	2.22	3.42	4.09
c3	2.32	2.08	1.60	2.95	3.47
c4	3.02	2.94	2.68	3.28	4.04
c5	2.55	2.38	1.62	2.67	3.71
c6	3.06	1.73	1.61	3.42	3.76
c7	1.70	2.42	1.91	2.28	3.32
c8	2.84	1.84	1.51	3.06	4.03
c9	1.90	1.73	1.27	2.33	3.20
c10	1.76	1.67	1.12	2.31	3.41
c11	1.99	2.67	1.58	2.86	4.22
c12	2.00	1.92	1.52	3.12	4.22
c13	2.06	2.70	1.51	3.15	4.01
c14	1.91	2.01	1.46	2.54	3.87
c15	1.12	1.55	1	1.90	2.74
c16	2.86	3.71	2.28	3.95	**4.92**
c17	2.22	2.74	1.68	3.55	3.71
c18	1.45	2.07	1.14	2.46	3.16
c19	1.82	2.29	1.32	2.95	3.50
c20	1.28	1.82	1.11	2.59	3.19
Level					
1	4.94	7.90	94.16	1.00	1
2	59.58	70.23	9.64	71.00	3.54
3	1	1	1	1.17	21.83
[total]	65.52	79.13	104.80	73.18	26.37

Cluster 2 : average users / Cluster 4 : highly addicted users
Other Clusters : potentially addicted users

According to the Table 3, highly addicted users have nearly 5 points for c1(How often do you find that you stay on-line longer than you intended?) and c16(How often do you find yourself saying "just a few more minutes" when on-line?). Two items are related to self-control. It means that the ability for self-control affects on the Internet overindulgence. If a person can control himself or herself, s/he can overcome from the addiction.

However, the testing item such as c15(How often do you feel preoccupied with the Internet when off-line, or fantasize about being on-line?) rarely affected to the respondents. It means that this item is not a good question when we classify the respondents into groups.

On the other hand, in the simple K-means analysis, the result is shown in Table 4. According to the simple K-means algorithms, only two clusters were

Table 4 Analysis with simple K-means algorithm

		Cluster#	
Attribute	Full Data	0	1
	(334)	(185)	(149)
c1	3.5928	3.1297	4.1678
c2	2.9401	2.4432	3.557
c3	2.2605	1.7297	2.9195
c4	3.024	2.8	3.302
c5	2.3353	1.9027	2.8725
c6	2.4371	1.8378	3.1812
c7	2.1647	1.9297	2.4564
c8	2.3353	1.7459	3.0671
c9	1.8473	1.4973	2.2819
c10	1.7725	1.3351	2.3154
c11	2.3593	1.827	3.0201
c12	2.2246	1.6595	2.9262
c13	2.4072	1.8486	3.1007
c14	2.0659	1.6162	2.6242
c15	1.4581	1.1405	1.8523
c16	3.2485	2.7135	3.9128
c17	2.5569	2.0216	3.2215
c18	1.8293	1.373	2.396
c19	2.1287	1.5838	2.8054
c20	1.7605	1.2324	2.4161

```
Clustered Instances

0       185 ( 55%)
1       149 ( 45%)
Class attribute: class
Classes to Clusters:

  0   1  <-- assigned to cluster
104   0 | 1
 81 128 | 2
  0  21 | 3
```

found. However, there exists the same result with EM algorithm. In case of cluster 1, the testing item, c1, is the highest point among all the items. In addition, c15 has the lowest point among all the items. If we assign a different weighting value to each items, we can examine adolescents' overindulgence level more correctly. In addition, based on the testing result, we can educate them not to addict to the Internet or advise them to prevent from the Internet addiction.

4 Results and Discussions

In this paper, we examine Young's Internet overindulgence testing method. From our experiment, we noticed that there are some problems in the method. Firstly, the assumption of independence among testing items can cause double counting bias. It can be proved from the factor analysis. Secondly, the results of Lisrel analysis showed that the measurement item independence assumption is rejected by chi-square difference test. It means that the method to classify the addiction level by summing the score of 20 testing items can be wrong.

On the other hand, we knew from clustering analysis, each addiction level has some unique features. The most remarkable conclusion is that the self-control factor is the most important in the Internet overindulgence. Thus, when we test adolescents' Internet overindulgence level, if we differentiate the testing items, then we can perform our test more correctly.

References

[1] Young, K.S.: Internet addication: The emergence of a new clinical disorder. Cyber Psychology and Behavior 1(3) (1996)
[2] http://www.netaddiction.com/resources/internet_addiction_test.htm
[3] Goldberg, I. K.: Internet addiction disorder (1996), http://willow.uml.edu/www/counseling/netdisorder.htm
[4] Davis, R.A.: A cognitive-behavioral model of pathological Internet use(PIU). Computers in Human Behavior 17 (2001)
[5] Kang, M.C., et al.: A study on Development of the Internet Addiction Metric. The Korean Journal on Adolescent Counseling Research 9 (2001)
[6] Ian Witten, H., et al.: Data Mining: Practical machine learning tools and techniques, 2nd edn. Morgan Kaufmann, San Francisco (2005)
[7] http://www.cs.waikato.ac.nz/~ml/weka/book.html
[8] Anderson, J.C., Gerbing, D.W.: Structural Equation Modeling in Practice: A Review and Recommended Two-Step Approach. Psychological Bulletin 103(3) (1988)

How Checkpointing Can Reduce Cost of Using Clouds?

Sangho Yi* and Derrick Kondo

INRIA Grenoble Rhone-Alpes, France
sangho.yi@inria.fr

Abstract. With the recent advent of Cloud Computing, the vision of computing as a utility has reached new heights. Resource prices can now change dynamically based on current demand and supply. In December 2009, Amazon released Spot Instances, which sell the spare capacity of their data centers. This spare capacity is sold using an auction-like bidding system. Instances are allocated or terminated depending on whether the current instance price is above or below the user's bid price. While pricing and reliability can vary dramatically and dynamically, current middleware have few mechanisms that deal with the cost-reliability trade-offs effectively. Also the failure rates are highly depending on the number of bids, the degree of bid, and the number of available resources. In this paper, we present how checkpointing can be used to reduce the cost and volatility of resource provisioning. Based on the real, recent price history of the spot instances we show the performance of the checkpointing schemes including the proposed one in terms of monetary cost and completion time of tasks.

1 Introduction

Recently, Cloud Computing has been the hottest topic in the world in both industry and academia. Clouds allow us to allocate and deallocate resources including computing power, and massive amount of disk space almost instantaneously and transparently on an as-need basis.

Price of the Cloud resources can differ in at least two ways. First prices can differ by vendor. The increasing number of Cloud Computing vendors has created a diverse market with different pricing models for cost-cutting, resource-hungry users. Second, prices can be changed dynamically (as frequently as an hourly basis) based on current demand and supply for the resources. In December 2009, Amazon released Spot Instances [1], which sell the spare capacity of their data centers. Their dynamic pricing model is based on users' bids. If a user's bid price is above the current Spot Instance price, the user can use the instance with the current price. If at any time the current price is above bid price, Amazon terminates the instance of the user. Thus, there is a trade-off between the cost of the instance and its reliability.

* Corresponding author.

M.-W. Han and J. Lee (Eds.): EKC 2010, SPPHY 138, pp. 85–95.
springerlink.com

While the computing market is liquid as ever, the current middleware run on top of these infrastructures currently cannot leverage changes in pricing or reliability. The middleware require mechanisms to seek by itself the cheapest source of computing resources given the demands of the application and current pricing.

In this paper, we present how checkpointing can be used to reduce costs and volatility of resources. Using the real, recent price traces of Amazon's Spot Instances, we show the performance of the checkpointing schemes including the proposed one and the previously made. Our main result is that the proposed checkpointing strategy significantly reduces the monetary cost, while improving reliability.

The remainder of this paper is organized as follows. Section 2 describes related work. Section 3 presents checkpointing strategies on Amazon EC2's spot instances. Section 4 evaluates performance of several checkpointing strategies based on the previous price history of the spot instances. Finally, Section 5 presents conclusions and possible extensions of this work.

2 Related Work

Cloud Computing vendors, such as Amazon Elastic Compute Cloud [1] and RackSpace Cloud Servers [2], provide ways of computing that broadly shares resources instead of using software and hardware resources on a local machine. In the Clouds vendors' points of view, maximizing utilization of the computing resources is the main issue to maximize profit. For this reason, recently Amazon started to use a bidding system for pricing the remaining computing resources, which is called Spot Instances [3]. The price of Spot Instances is much less expensive than the normal instances, but users may have unavailable durations due to the number of bidders, available resources, and the amount of bid. RightScale [4] is a third party cloud computing broker which gives cloud management services for Cloud users. They provide several added features for users to reduce complexity of managing the Cloud resources. However, they still do not have any service for efficiently utilizing the spot instances on Amazon EC2. Thus the users of the spot instances have to take into account the failures, and manage their tasks with some fault-tolerance mechanisms.

There are several challenges related to checkpointing on the unreliable resources. The first one is finding the relationship between past and future failures for proactive task migration mechanisms. Much work was done for understanding dependence between failure events [5, 6]. Another challenge is using an efficient checkpointing method for minimizing the expected execution time in the presence of failures. This also has been the subject of previous work described in [7, 8, 9].

Checkpointing is a fault-resilience mechanism which takes checkpoints at certain points, and rollbacks to the latest checkpoint when recovering from a failure. For several decades researchers have investigated efficiency of many kinds of checkpointing methods, including scheduling of checkpoints. In [7], A. Duda proved the optimal placement of checkpoint if the cost is constant. In [8], Yi et al. proposed an adaptive checkpointing scheme which adaptively decides efficient points of checkpoints when the cost of checkpointing changes over time. In [10], we studied the performance of several existing checkpointing schemes in the Amazon's Spot Instances. The results showed that an appropriate checkpointing strategy can reduce monetary costs while increasing reliability of the volatile resources.

3 Spot Instances on Amazon Ec2

In this section we describe the system model used in this paper and introduce the considered checkpointing schemes.

3.1 System Model

Amazon allows users to bid on unused EC2 capacity provided as 64 types of Spot Instances that differ by computing / memory capacity, system software, and geographical location [3]. Their prices called Spot Prices change dynamically based on supply and demand for the resources.

Figure 1 shows examples of spot price fluctuations for three ***eu-west-1.linux*** instance types during 8 days in the beginning of June, 2010. Customers whose bids meet or exceed the current spot price gain access to the requested resources. The following system model was made according to the characteristics of Amazon EC2's spot instances.

- Amazon provides a spot instance when a user's bid is greater than the current price.
- Amazon stops instances without any notice when a user's bid is less than or equal to the current price.
- Amazon does not charge the latest partial hour when Amazon stops an instance.
- Amazon charges the last partial hour when a user terminates an instance.
- The price of a partial-hour is considered the same as a full-hour.
- Amazon charges each hour by the last price.
- Amazon freely provides the spot price history.
- The price of storage service is almost negligible.

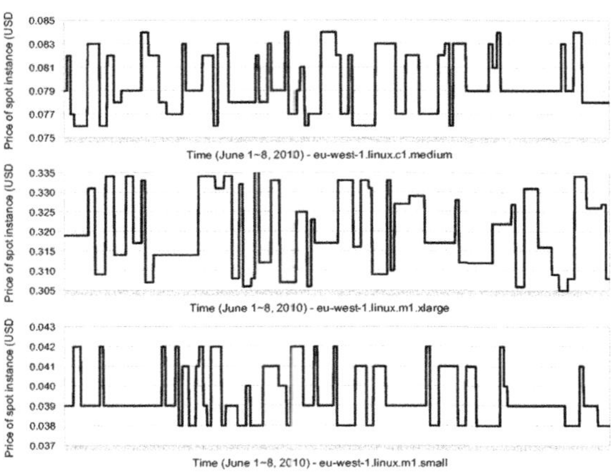

Fig. 1 Spot price fluctuations of *eu-west-1.linux* instance types

Figure 2 shows an example of probability density function of failures on ***eu-west-1.linux.c1.medium*** type from January 1st to June 24th, 2010. This graph shows the difference of failure rate according to the different user's bid price.

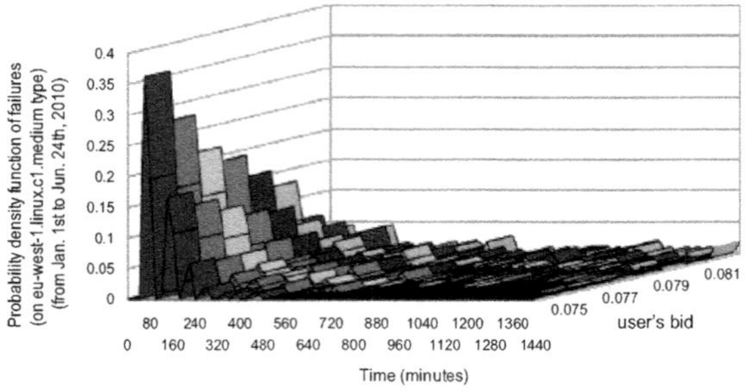

Fig. 2 Probability density function of failures on *eu-west-1.linux.c1.medium* instance type

3.2 Previous Checkpointing Schmes

In the previous work [10], we compared performance of 12 checkpointing schemes on the real traces of spot prices. In the results, we found that the existing checkpointing schemes cost almost 50% higher than the optimal case of checkpointing, and even the best one among the schemes costs 30% higher than the optimal case. Also, we have found that the hourly checkpointing[1] shows reasonable performance in terms of the monetary costs because Amazon does not charge the latest partial hour when Amazon stops an instance. But, sometimes non-hourly checkpointing shows better results than hourly checkpointing, because even the partial-hours stopped by Amazon can be used for the execution because the checkpoint can be taken just before the failure. Thus, we need better checkpointing strategy that aware of the expected occurrence of failures and the characteristics of the Amazon EC2's Spot Instances.

3.3 Proposed Checkpointing Scheme

In this section, we propose a checkpointing scheme called EKC. It stands for an efficient, knowledge-based checkpointing for the Spot Instances. EKC is mainly based on the adaptive checkpointing interval determination mechanism [8,10], and uses some known things from Amazon's Spot Instances. Listed below are the known characteristics of the Spot Instances.

- The probability distribution of failure occurrences depend on the current spot price.
- Amazon does not change the spot price more than once per hour.

---

[1] Hourly checkpointing scheme takes checkpoints at the end of every full-hour.

- Users can delay termination of running instances up to the end of hour-boundary, because when Amazon stops the instances the users do not pay the last partial-hour.

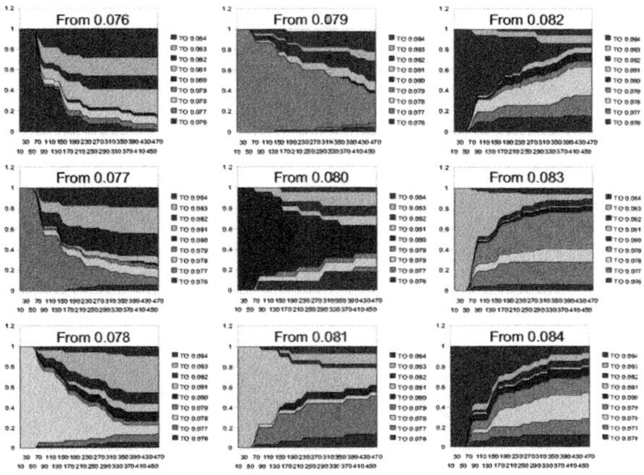

Fig. 3 Probability distributions of price change of eu-west-1.c1.medium instance type

Figure 3 shows the probability distribution of spot price change of ***eu-west-1.c1.medium*** according to the current spot price and the time duration. This result shows the fact that the probability distribution is not uniform according to the current spot price. The low current prices (e.g. From 0.076~0.077) have higher transition probability to the high prices (To 0.081~0.084). On the other hand, if the current price is in the middle (From 0.079), there's less transition probability to the high prices. Thus, the checkpointing scheme can utilize the different probability distribution information to get better prediction of upcoming failures.

We substituted the probability density function of failure $f(k,u_b)$ with $f(k,p_c,u_b)$ from the Theorem 1, 2, and 3 in [10], where p_c is the current spot price. The following are the modified equations of Theorems used in this paper.

Equation 1. The expected execution time $T(t)$ of a process without checkpointing when executing t time units is

$$T(t) = \frac{t + \sum_{k=0}^{t-1} (k+r-t) f(k,p_c,u_b)}{1 - \sum_{k=0}^{t-1} f(k,p_c,u_b)} \quad (1)$$

Equation 2. The expected recovery time when skipping a checkpoint at t time units after taking checkpoint, $R_{skip}(t)$ is given by

$$R_{skip}(t)=\sum_{k=0}^{t_r-1}\left(k+r+T(t)\right)f\left(k,p_c,u_b\right) \tag{2}$$

Equation 3. The expected recovery time when taking an hour-boundary checkpoint at t time units after taking checkpoint, $R_{take}(t)$ is given by

$$R_{take}(t)=\begin{array}{l}\sum_{k=0}^{t_r-1}(k+r)f\left(k,p_c,u_b\right)\\+\sum_{k=0}^{t_c-1}T(t)f\left(k,p_c,u_b\right)+T\left(t_c\right)\end{array} \tag{3}$$

We skip to show their proofs because our substitution from $f\left(k,u_b\right)$ to $f\left(k,p_c,u_b\right)$ does not affect the way of proof of the equations. See [10] for details.

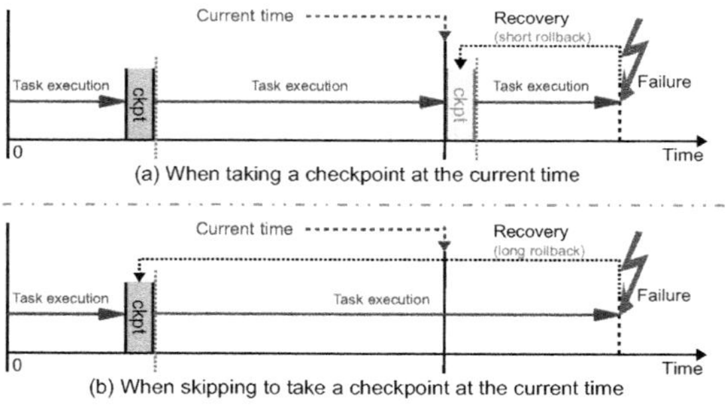

Fig. 4 Effects of skipping and taking a checkpoint on the recovery cost

Figure 4 shows the effects of the checkpointing on the recovery cost. When we do not take a checkpoint at a certain point, we should have long rollback when failure occurs in the future. Otherwise, if we take a checkpoint, we can recover with less cost.

Our proposing checkpointing strategy, namely EKC, works based on the basic known characteristics of the Spot Instances. For example, when the current spot price has changed, then we will not have fluctuations on the price up to 1 hour. Thus, we don't need to take a checkpoint up to the end of hour-boundary from the beginning point of the current spot price. Otherwise, EKC makes decision whether to take or skip a checkpoint every certain minutes, and uses adjustable sliding window to calculate $f(k, p_c, u_b)$ based on the spot price history. The bigger sliding window is better for the long-term expectation, but it may lose some recent behaviors of the spot price fluctuations.

Based on the equations, we can decide which option can reduce the recovery cost in the presence of failures. In more detail, our policy takes a checkpoint when $R_{skip}(t) > R_{take}(t)$, and skips it otherwise.

4 Performance Evaluation

In this section, we show the impact of checkpointing policies in terms of the monetary costs and execution time on some instance types in Amazon EC2.

Table 1 Simulation parameters used in this paper

Parameter	Value
Starting date of past traces	Jan. 11th, 2010
Ending date of past traces	Jun. 24Th, 2010
Past traces (for calculating pdf of failure)	43,200 minutes
Minimum bidding granularity	0.001 USD
Work requirement of a sample task	1,000 minutes
Number of simulation for each point	100 times
Checkpointing cost (in time)	5 minutes
Restarting (rollback) cost (in time)	10 minutes

Table 1 shows the simulation parameters in detail. We assume that the applications can expect the cost of checkpointing[2]. We assume that the total work requirement of each program is 1,000 minutes, and we used the latest 30 days (43,200 minutes) of price history to get the probability density function of failure occurrence $f(k, p_c, u_b)$. Our EKC scheme makes a decision whether to take or skip a checkpoint every 10 minutes. We will focus on the impact of their parameters on the monetary costs and execution time in our future work.

We have selected 6 different checkpointing policies from the previous work [10], and compared the performance of them with our proposed scheme, EKC. Listed below are the abbreviation and their meaning.

2 Using variable checkpointing cost is also possible in our system model. We will show the case of variable checkpointing cost which is depending on the size of tasks in our future work.

- OPT: It takes checkpoints just before the failures.
- NONE: It does not take checkpoints.
- HOUR: It takes checkpoints every hour.
- EDGE: It takes checkpoints when price increases.
- AF(10): It determines whether to take or skip a checkpoint every 10 minutes based on the probability density function of failures, $f\left(k,u_b\right)$.
- AF(30): This is the same as AF(10) except for the decision making period (30 minutes).
- EKC: It makes decision every 10 minutes whether to take or skip a checkpoint based on the yet-another probability density function of failures according to the current price, $f\left(k,p_c,u_b\right)$.

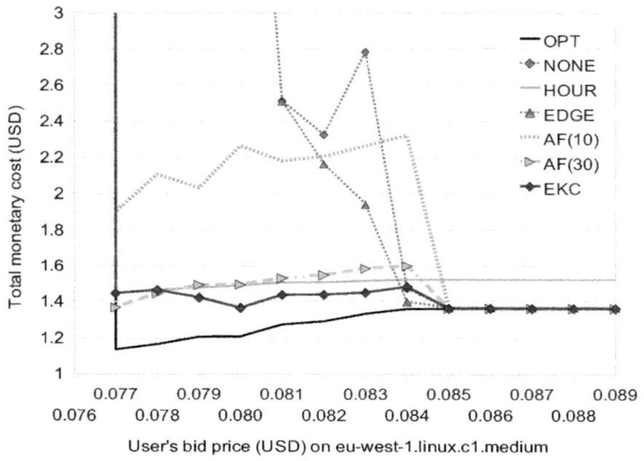

Fig. 5 Total monetary cost of execution on *eu-west-1.linux.c1.medium* instance type

We have picked the ***eu-west-1.linux.c1.medium*** as a representative instance type to evaluate the total price of a task, its completion time, and a product of both as a combined metric. Figure 5 shows the total monetary cost of execution of a sample task on the investigated instance type. Obviously EDGE performs poorly, while HOUR shows reasonable performance. EKC shows the closest performance to OPT for most of user's possible bid prices.

Figure 6 shows the total execution time on the same instance type. In this result, EKC shows the best for all policies (except for OPT).

Figure 7 shows the normalized results of the Figs. 5 and 6. Based on the results, we can observe that the proposed scheme called EKC has almost less than 15% overhead compared to OPT in terms of both the monetary cost and the execution time, while the previous schemes have more overhead than that of the proposed scheme.

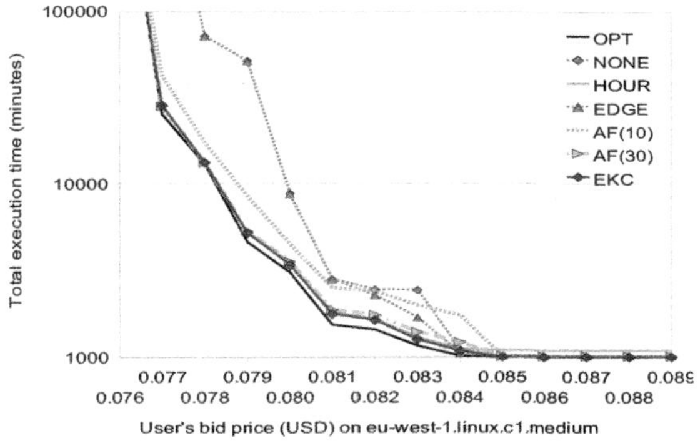

Fig. 6 Total execution (task completion) time on *eu-west-1.linux.c1.medium* instance type

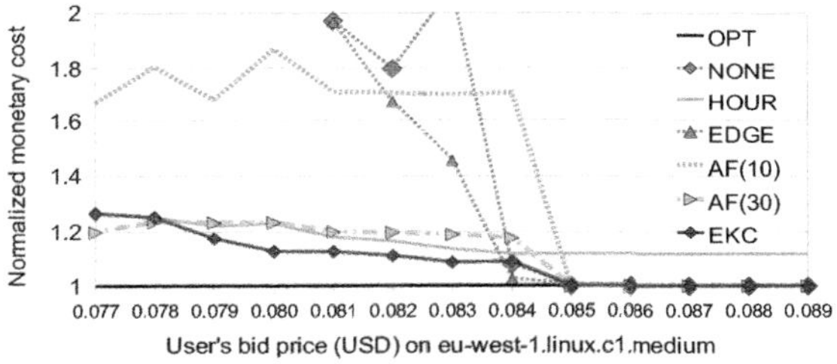

(a) Normalized monetary cost

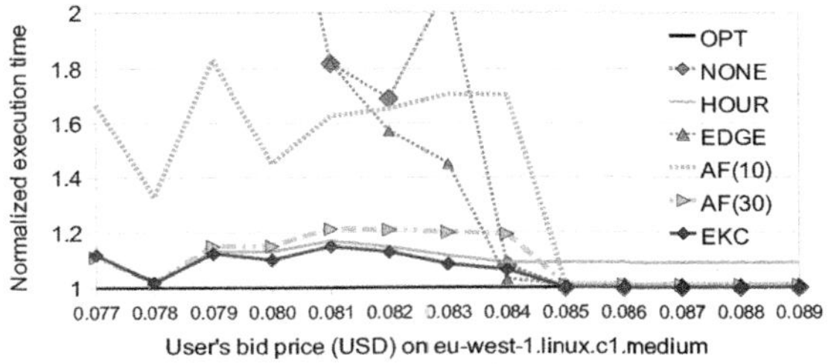

(b) Normalized execution time

Fig. 7 Normalized results on *eu-west-1.linux.c1.medium* instance type

Figure 8 shows the normalized 'product of the monetary cost and the execution time'. From this, we observe that EKC outperforms all the other schemes in most cases, while other schemes show higher overhead than the optimal case.

We made our simulation for all the other instances on eu-west zone. The results are available on the web (see Section 6). Taking simulations on another zones or instances is also possible by using the recent traces of spot instances.

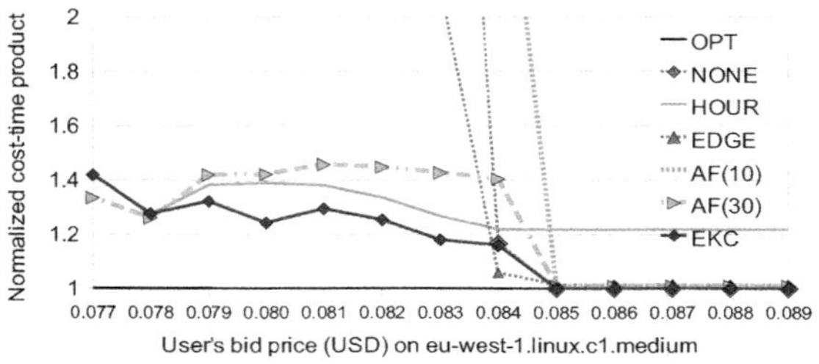

Fig. 8 Normalized product of monetary cost and execution time on *eu-west-1.linux.c1.medium* instance type

From the results, we convinced that our proposed scheme, EKC can reduce both the monetary cost and the execution time compared with the existing schemes.

5 Conclusions and Future Work

We proposed a checkpointing strategy called EKC, which uses both the adaptive on-line decision mechanism and some known characteristics of Amazon's Spot Instances. We showed the performance of EKC and the previous checkpointing mechanisms including the optimal case based on the price history given by Amazon. Our evaluation results show that the EKC outperforms the previous checkpointing mechanisms in terms of both the monetary costs and the execution time.

We are currently extending our work to design a Cloud resource brokering solution based on the checkpointing mechanism [11]. We are also interested in developing better prediction mechanism to maximize the efficiency of checkpointing under volatile Cloud resources.

6 Availability

All data used in this study, the full source code of the simulator and additional results are available under the following web URL: http://spotckpt.sourceforge.net

References

[1] Amazon EC2 web, `http://aws.amazon.com/ec2/`
[2] RackSpace web, `http://www.rackspace.com/`
[3] Amazon EC2's Spot Instances web, `http://aws.amazon.com/ec2/spot-instances/`
[4] Right Scale web, `http://www.rightscale.com/`
[5] Fu, S., Xu, C.-Z.: Exploring event correlation for failure prediction in coalitions of clusters. ACM/IEEE Supercomputing Conference (SC), pp. 1–12 (September 2007)

[6] Javadi, B., Kondo, D., Vincent, J., Anderson, D.P.: Mining for Availability Models in Large-Scale Distributed Systems: A Case Study of SETI@home. In: IEEE/ACM International Symposium on Modelling, Analysis and Simulation of Computer and Telecommunication Systems, MASCOTS (September 2009)
[7] Duda, A.: The effects of checkpointing on program execution time. Information Processing Letters 16(1), 221–229 (1983)
[8] Yi, S., Heo, J., Cho, Y., Hong, J.: Taking point decision mechanism for page-level incremental checkpointing based on cost analysis of process execution time. Journal of Information Science and Engineering 23(5), 1325–1337 (2007)
[9] Plank, J.S., Li, K., Puening, M.A.: Diskless Checkpointing. IEEE Transactions on Parallel and Distributed Systems 9(10), 972–986 (1998)
[10] Yi, S., Kondo, D., Andrzejak, A.: Reducing Costs of Spot Instances via Checkpointing in the Amazon Elastic Compute Cloud. In: Proceedings of the 3rd International Conference on Cloud Computing (July 2010)
[11] Andrzejak, A., Yi, S., Kondo, D.: Decision Model for Cloud Computing under SLA Constraints. In: MASCOTS 2010 (August 2010)

Cost Effective High-Voltage IC Technology Implemented in a Standard CMOS Process

Jong Mun Park*, Rainer Minixhofer, and Martin Schrems

austriamicrosystems AG, Schross Premstaetten, Austria
Jong-mun.park@austriamicrosystems.com

Abstract. For competitive high-voltage (HV) integrated circuit (IC) products an excellent trade-off between specific on-resistance $R_{on,sp}$ and breakdown voltage BV of a HV lateral DMOS (LDMOS) transistor, while keeping low fabrication cost, is mandatory. This paper presents a review of the HVIC technology trend with special emphasis on cost effective 0.35 μm and 0.18 μm HV-CMOS technologies. Through optimized process setup and device engineering a very competitive $R_{on,sp}$-BV trade-off of a HV LDMOS transistor without degrading the low-voltage (LV) CMOS performance has been achieved. A 0.35μm HV-CMOS technology with LDMOS transistor operating voltages from 20V to 120V is reported. Only two mask level adders on top of standard CMOS are required to provide the full set of 3.3V, 5V and 20V-120V HV devices. This is the result of taking advantage of predictive TCAD which enables early optimization of device layouts and dopant concentrations. In addition, HV and LV process integration issues of a 0.18 μm HV-CMOS technology, which play a key role to efficiently implement a HV module into a deep submicron CMOS process, are described. Key issues of p-channel LDMOS transistors are reviewed. The hot-carrier (HC) behaviour of a 50 V p-channel LDMOS transistor is presented too.

1 Introduction

In this publication any device that can operate at voltages higher than 5 V is considered a high-voltage (HV) device. There is a broad spectrum of applications for HVICs in consumer, communications, industrial, and automotive electronics, including ICs, as diverse as display drivers, motor controllers, power management ICs, data-line drivers, and even MEMS actuator drivers.

HVIC process technologies maybe divided into two classes. Each class is focused on different application regimes. The first class, which we will name HV-CMOS technologies, provides HV devices as an extension to a CMOS base process by adding additional implantation steps, a thicker HV gate oxide, and by using layout measures such as drain extensions. HV-CMOS processes are typically focused on applications with HV device currents below 300 mA and operating voltages typically centered around 20 V and 40 V with maximum operating voltages below 100 V. Due to their simplicity these processes are suitable for cost-sensitive large volume consumer applications like display driver ICs. The second

* Corresponding author.

M.-W. Han and J. Lee (Eds.): EKC 2010, SPPHY 138, pp. 97–105.
springerlink.com

class of processes called BCD (Bipolar-CMOS-DMOS) introduces n- and p-type buried layers in order to provide HV devices with lower specific on-resistance $R_{on,sp}$ and better isolation to substrate as compared to HV-CMOS processes. It has been recognized recently that the buried layer limits the voltage scalability leading to HV device architectures with pulled back buried layers which makes the process architecture more similar to HV-CMOS processes.

In this paper we will give report on a 0.35μm HV-CMOS technology which provides similar $R_{on,sp}$ levels as BCD technology, and thus entering an application regime previously reserved for BCD technologies only. On-resistances as low as 0.25 $m\Omega cm^2$, 1.1 $m\Omega cm^2$, and 4.4 $m\Omega cm^2$ were achieved for NMOS devices with a breakdown voltage (BV) of 30 V, 50 V and 150 V, respectively. Up to now only much more complex BCD processes could achieve such low on-resistance values. An ESD (electro-static discharge) capability up to 8kV HBM (human body model) and a set of highly accurate device models enable complex power IC designs such as one of the first fully functional Flexray Transceivers recently qualified by two large European car manufactures.

A HV-CMOS technology implemented in a standard 180 nm RF-CMOS platform is reviewed in this paper too. It includes n- and p-channel LDMOS transistors with BV over 80 V. The measured output characteristics of HV-LDMOS transistors show excellent $R_{on,sp}$-BV trade-off (e.g. p-channel LDMOS transistors: BV=-80~-85 V, $R_{on,sp}$=1.6~1.8 $m\Omega cm^2$). This new HVIC technology enables the design of highly integrated SoC (system-on-chip) applications like lighting management units. The combination of lowest power consumption with lowest chip cost is required for any high volume device. By using the high gate-density of 0.18 μm CMOS in combination with the low on-resistance of HV devices for 5 V, 20 V and 50 V the reported technology is very attractive for a large variety of products.

2 High-Voltage CMOS Technology

2.1 Technology Overview

HVIC have always used design rules and technologies which are less aggressive than that used for ULSI and VLSI devices. The strong drive towards integration of HV devices and LV-CMOS leads to single chip systems for low power applications.

Recently, technology node for HVIC went down from 0.35 to 0.18 μm and further down to 0.13 μm, which offers the possibility of integration of high density digital IP together with HV devices. However, for the HV areas the small ground-rules do contribute 1:1 to a reduction the device area, because in many cases most of the chip size is determined by the HV device area, which scale far less with ground-rules than LV-CMOS. In addition, it is not trivial to implement HV devices and LV-CMOS into a single chip monolithically while keeping excellent device performance for both HV devices and LV-CMOS. Furthermore, manufacturing costs must be considered already during the development of HVIC technology. Many new device concepts have been proposed to improve the performance of HV

LDMOS transistor. As an example, Fig. 1 shows the 150 V super-junction LDMOS transistor on SOI (silicon- on-insulator). Such a new structure greatly helps to reduce the on-resistance of the device, although the fabrication process is very complicated.

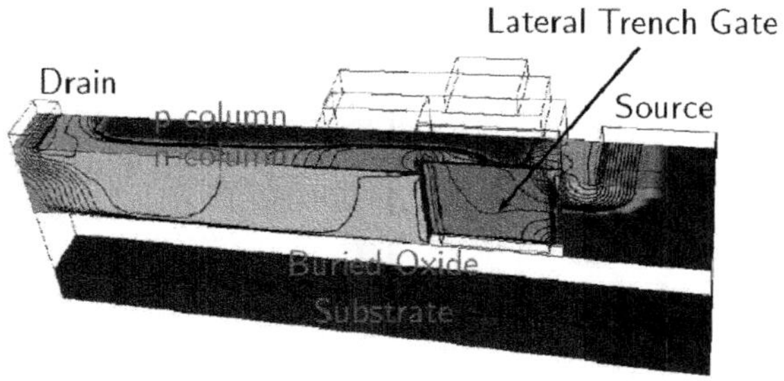

Fig. 1 150 V n-channel super-junction SOI-LDMOS transistor with lateral trench gate [1]

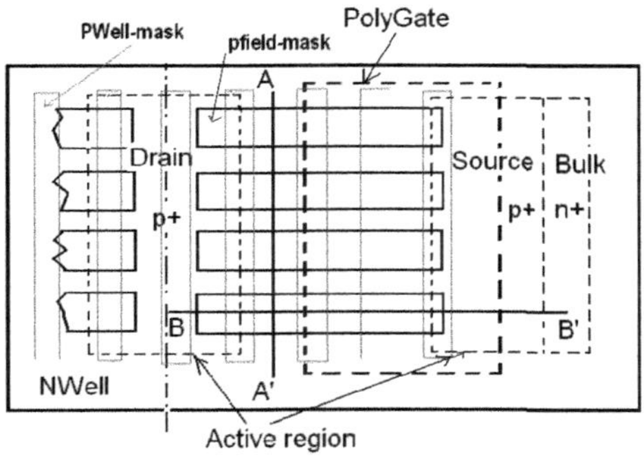

(a) Layout (top view).

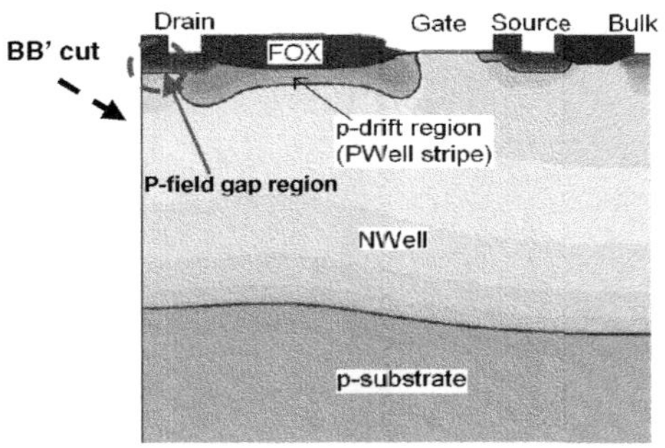

(b) Structure (BB' cut).

Fig. 2 20 V p-channel LDMOS transistor in a standard 0.35 μm CMOS technology [2]

Another approach is to introduce HV transistors in a standard CMOS process without any additional mask steps. However, NWell and PWell implant doses in a standard CMOS process are quite high, and they cannot be applied directly to form the drift region or body of HV transistors because of premature breakdown or punch-through. Fig. 2 shows an HV p-channel LDMOS transistor implemented in a standard CMOS process without any additional mask steps, which has a BV over 20 V. The p-channel LDMOS transistor shows a p-drift region similar to a super-junction structure. Such a structure is introduced under the field oxide (FOX) in the NWell. Carefully controlled PWell stripes together with p-field implant and anti punch-through implant are used to form the super-junction like structure. No additional process steps are necessary for this approach.

2.2 0.35 µm HV-CMOS Technology

The 0.35 µm HV-CMOS technology (Fig. 3) was developed as a fully modular extension of an industry-standard two-poly and four-metal (2P4M) 0.35µm CMOS process with 3.3V and optional 5V gates. Therefore the gate density of 17000 gates/mm^2 could be maintained without compromises. 20 V to 120 V LDMOS transistors have been realized by adding a set of NWell and PWell and an optional 50 nm gate oxide supporting gate voltages of 20 V. n- and p-channel LDMOS transistors with operating voltages such as 20 V, 50 V, 70 V, and 120 V were provided without any process changes by optimizing device layouts only. In addition to 3.3 V, 5 V low voltage devices and 20 V~120 V HV devices, the technology also offers large set of complementary devices such as lateral bipolar devices, JFETs, resistors up to 1.2 kOhm and PIP (Poly Insulator Poly) capacitors.

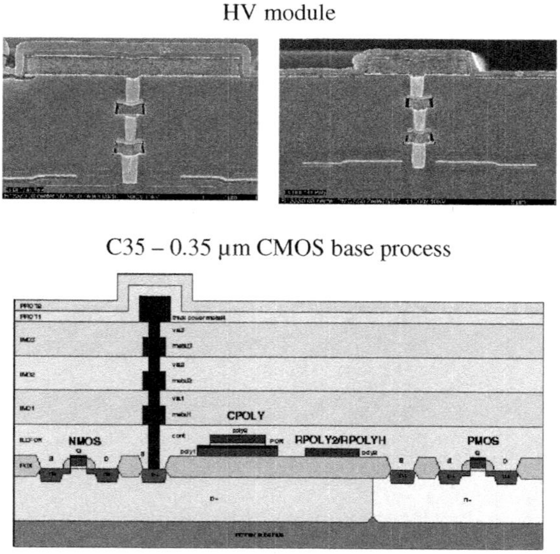

Fig. 3 0.35 µm HV-CMOS technology concept

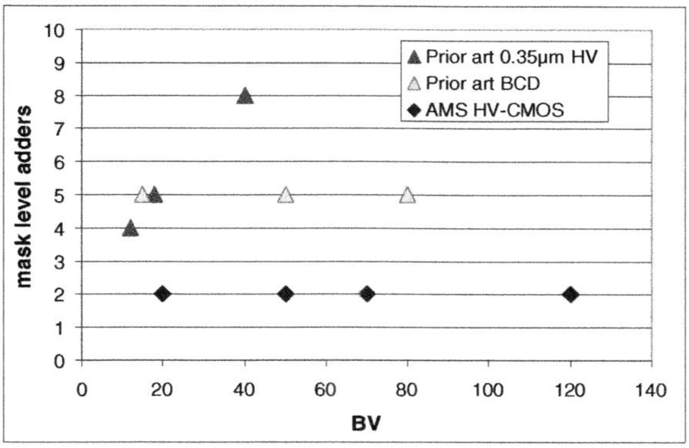

Fig. 4 Mask level adders

The technology only requires two mask level adders on top of the CMOS base process. This is the lowest number of mask level adders reported for any technology providing up to 120 V devices and leads to lowest reported process complexity (Fig. 4). $R_{on,sp}$ were measured from the output characteristics of the LDMOS devices at a gate voltage of V_{GS}=0.1V. $R_{on,sp}$ values of 25 mΩmm^2, 110 mΩmm^2, 140 mΩmm^2 and 440 mΩmm^2 were achieved for n-channel LDMOS transistors with operating voltages of 20 V, 50 V, 70 V, and 120 V as well as BV of 30 V, 70 V, 90 V, and 150 V p-channel LDMOS $R_{on,sp}$ values were typically two times higher than for n-channel LDMOS which can be explained by the differences in majority carrier mobility.

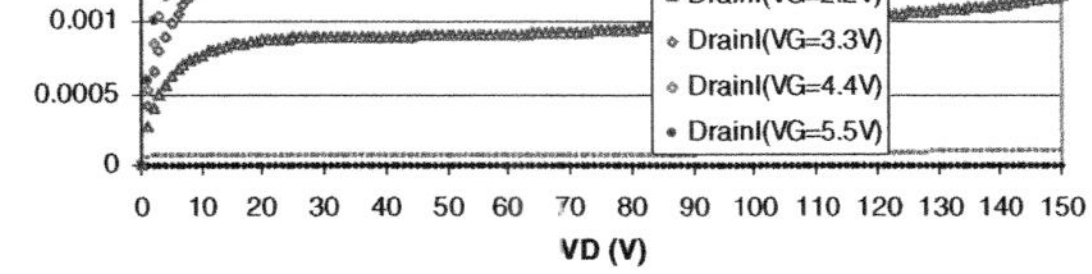

Fig. 5 I-V curve of a 120 V n-LDMOS (V_{Gmax}=5.5 V)

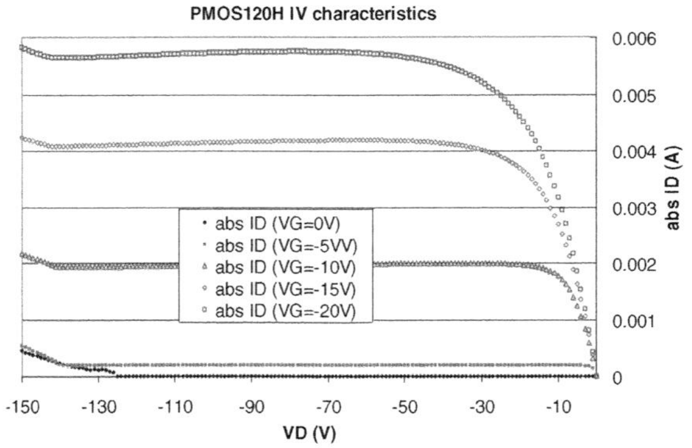

Fig. 6 I-V curve of a 120 V p-LDMOS (V_{Gmax}=20 V)

The measured output characteristics for a 120 V n-channel LDMOS with V_{GS}=5.5 V and a 120 V p- channel LDMOS with V_{GS}=20 V are shown in Fig. 5 and Fig. 6, respectively. As a final verification of the robustness of this technology a Flexray bus transceiver product has been developed (Fig. 7). The product could already meet the specifications laid out by the Flexray standard after the first design verification run. One critical parameter in the specification of the Flexray protocol is achieving certain limits for the differential voltages between the high side bus and the low side bus. There is a good correspondence between measurements and simulations for both the mean and the standard deviation.

Parameter	min.	max.
Battery Supply Voltage	-0.3V	+50V
DC Voltage	-50V	+50V
Junction Temp. TJ	-40C	+150C
ESD (HBM)	-4kV	+4kV
Latchup immunity	-100mA	100mA

Fig. 7 Flexray bus transceiver manufactured in the 0.35 μm 50V HV-CMOS technology

2.3 0.18 μm HV-CMOS Technology

The reported 0.18 μm HV-CMOS process is based on a standard 0.18 μm RF-CMOS technology platform and integrates fully compatible 1.8 V and 5.0 V LV-CMOS transistors. To reduce the number of masks, well sharing between LV and HV devices is the method of choice. The STI depth, corner rounding and oxide-thickness of the STI are determined mainly by design rules of the LV devices. However, the STI region of a HV-LDMOS transistor plays an important role to the device performance, because it is located in the drift-region and is thus an integral part of the active device region. The 1.8 V and 5V CMOS transistors and ground-rules are identical to the devices and ground rules of the existing RF-CMOS technology, which offers excellent 6T SRAMs (3.6 μm^2 per cell) and high gate-density (118 kgates/mm^2) with non-epi p^--substrate as a starting material. In addition, LV- CMOS transistors are isolated from the p-substrate by the DNWell. Table 1 lists the summary of key LV and HV devices. Just the HV LDMOS transistors with thick-GOX (Gate operating voltage of 20 V) are listed in the table, although the process offers p- and n-channel asymmetric LDMOS transistors (20 V, 25 V, and 50 V on the Drain) with thin- and mid-GOX (1.8 V, 5 V operating voltage at the Gate) too.

Table 1 Key LV and HV devices in the 0.18 μm HV-CMOS technology [3]

Devices	$V_{th,lin}$ [V]	Ids [μA/μm]	BV [V]	$R_{on,sp}$ [mΩcm^2]
1.8V LV NFET/PFET	0.35/0.42	600/260		
1.8V high V_t NFET/PFET	0.52/0.52	500/210		
5V LV NFET/PFET	0.62/0.6	630/345		
20V/50V NMOS	2.0/2.1		33/96	0.29/1.3
20V/50V PMOS	2.5/2.5		45/85	0.63/1.8
20V/50V symmetric NMOS	2.1/2.1	340/230	40/97	0.47/3.16
20V/50V symmetric PMOS	1.8/1.8	250/160	31/85	1.29/6.3
VNPN/VPNP	Gain = 50/70		30/100	
JFET	1.5		30	
MIM Cap.	2, 2.75, 4, and 5.5 fF/μm^2			
HV Cap.	0.244 fF/μm^2			

The HV n- and p-channel LDMOS transistors show excellent $R_{on,sp}$-BV trade-off. For the mid-GOX 50V NMOS $R_{on,sp}$ = 1.3 mΩcm^2 at BV = 96 V has been achieved while $R_{on,sp}$ and BV of a thick-GOX PMOS (50 V p-channel LDMOS transistor with V_{Gmax} = -20 V) were 1.8 mΩcm^2 and 85 V, respectively. Fig. 8 shows the comparison of $R_{on,sp}$-BV trade-off of p-channel LDMOS transistors [4,5]. The 0.18 μm HV-CMOS technology shows the best reported $R_{on,sp}$-BV trade-off for p-channel LDMOS transistors.

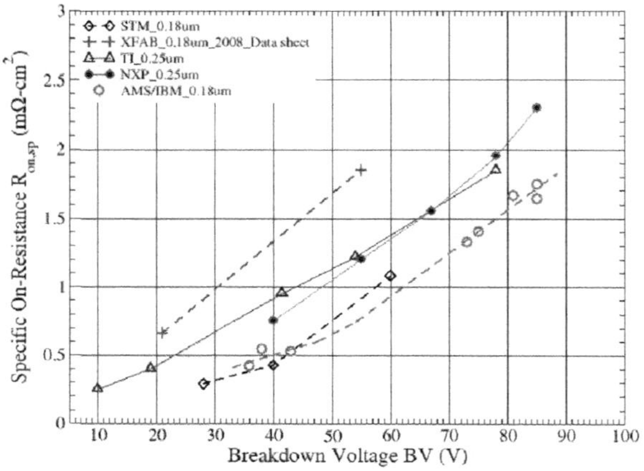

Fig. 8 R_{on}-BV trade-off of p-channel LDMOS transistors [3]

High electric-fields in a HVIC generate hot carriers, which cause device degradation by interface trap generation and charge trapping in the oxide. The peak electric-field at the STI corner in the drift- region is believed to be a major driver of the HC generation. Fig. 9 shows the potential distribution at BV (-99 V) and impact-ionization (V_D=-50 V and V_G=-1.5 V) at the STI corner of a p-channel LDMOS transistor. One can see clearly the strong impact-ionization at the sharp STI corner. Fig. 10 shows the linear-region of drain current Idlin degradation at the stress bias of V_D=-50 V and V_G=-1.5 V. As shown in the figure, p-drift implant dose and layout geometry affect the HC behaviour of the device.

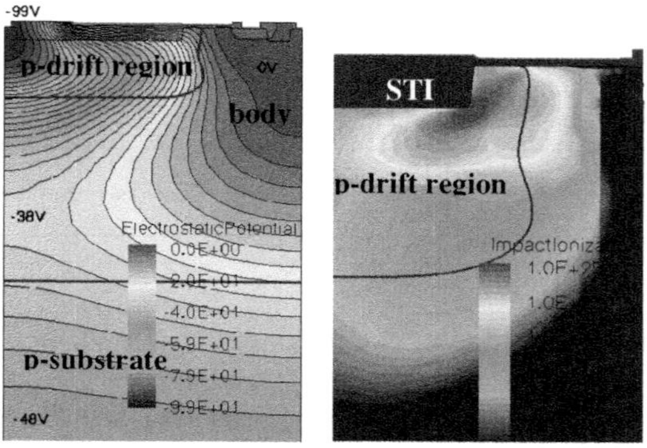

(a) Potential distribution at BV (b) Impact-ionization (II)

Fig. 9 p-channel LDMOS transistor. (a) Potential distribution at BV, and (b) II at V_D= -50 V and V_G=-1.5 V

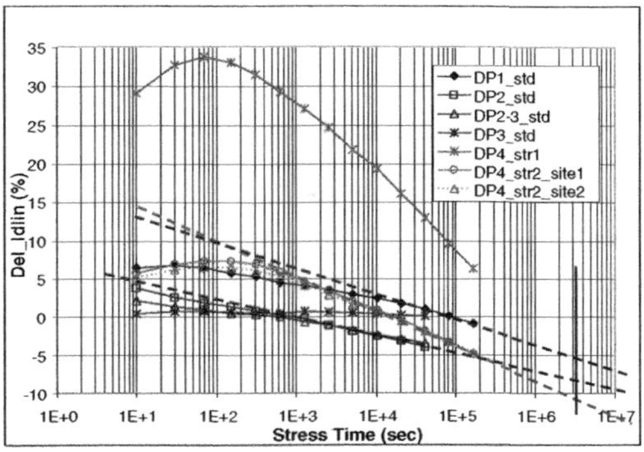

Fig. 10 Percent change of Idlin of p-channel LDMOS transistors. Drift-dose (DP1~DP4) and layout geometry (std:standard, str1, and str2) affect the HC behaviour.

3 Results and Discussions

The reported technology is the first 0.35μm HVIC technology reported that can cover a voltage regime up to 120 V and beyond. The proposed HV-CMOS technology requires only two mask level adders on top of CMOS while reaching the low on resistance values of complex BCD processes. Complicated HV designs such as a Flexray bus transceiver have been realized in H35. In addition, a state of the art of the HVIC technology implemented in a standard 0.18 μm CMOS process was described. The discussed HV-CMOS technology shows the best reported $R_{on,sp}$-BV trade-off for p-channel LDMOS transistors. The reported high performance 0.18 μm mixed-signal RF-CMOS/HV- CMOS technology offers a low cost and an efficient SoC design environment for a wide application range.

Acknowledgment. The authors wish to thank Dr. N. Feilchenfeld (IBM) and austriamicrosystems-IBM 0.18 μm HV-CMOS joint- development team for their great cooperation works.

References

[1] Park, J.M., et al.: ESSDERC, p. 96 (2002)
[2] Vescoli, V., et al.: ISPSD (2007)
[3] Park, J.M., et al.: ICEE, p.100 (2010)
[4] Riccardi, D., et al.: ISPSD, p. 73 (2007)
[5] Letavic, T., et al.: ISPSD, p. 60 (2008)

Breast Cancer Related Facts: Comparison between Korea and Austria

Hanna Sheu[1,*] and Ali Tahamtani Omran[2]

[1] Brustgesundheitszentrum, Departement of Surgery,
Krankenhaus der Barmherzigen Schwestern, Linz, Austria
hanna.sheu@gmail.com, hanna.sheu@bhs.at
[2] IT-Departement, Krankenhaus der Barmherzigen Schwestern,
Linz, Austria

Abstract. Background: Epidemiological data show that breast cancer is the most common cancer in women worldwide [1]. In Korea breast cancer still ranks behind stomach, lung, colorectal, liver, gallbladder cancer as a cause of death in women. Epidemiological features suggest that the breast cancer incidence rate in Korea will increase [2]. Although the incidence of malign breast tumor in Korea is significantly lower compared to western countries, the rising number of breast cancer patients in Korea is alarming [3]. Austria is one of the leading nations in breast cancer research [4].

Objective: This Summary of breast cancer related epidemiological facts and studies should provide information to a broad collective of scientists of different specializations at the Europe Korea Conference 2010. This paper should line out the difference of breast cancer related epidemiological facts between Korea and Austria and give basis for discussion concerning public health issues. Methods: Online research using Pubmed and Google as search tools relevant papers and statistics were collected and summarized.

Conclusion: The comparison of the epidemiological data indicates that the breast cancer incidence rate in Korea will reach the incidence rate in Austria in the future. Since Austria plays a leading role in the fight against breast cancer, the Austrian breast care standards could play a role model for the benefit of Korean public health.

1 Introduction

Breast cancer is the most common cause of death for women in Austria [5]. Since malign breast tumors are a major health burden in Austria, there are many efforts to investigate for the best evidence based therapy strategies. Austrian Breast and Colorectal Cancer Study Group generates valid and reliable clinical data by prospectively randomized clinical investigations. Since 1984 a total of 17.000 patients

* Corresponding author.

M.-W. Han and J. Lee (Eds.): EKC 2010, SPPHY 138, pp. 107–112.
springerlink.com

have been enrolled in their studies. The Austrian Breast and Colorectal Cancer Study Group is currently recruiting up to 40 per cent of all Austrian breast cancer patients to its clinical trial [4].

In Korea breast cancer still ranks behind stomach, lung, colorectal, liver, gall-bladder cancer as a cause of death in women. There are trends of increase of the incidence of breast cancer in Korean women [2]. This paper overviews the significant differences in epidemiological facts about breast cancer between Korea and Austria and discusses the possible prevention methods to decrease the mortality of breast cancer in Korea.

2 Materials and Methods

Since 1980 the cancer related epidemiological data is collected in a hospital based cancer registry called the Korea Central Cancer Registry (KCCR). Austria provides cancer statistics online by Statistik Austria.

For further online research Pubmed and Google were used as search tools.

Since the recent Korean breast cancer related data online was from 2005, the comparison was made for the incidence and mortality in 2005.

3 Results and Discussions

Results

Incidence and Mortality

In 2005, 9842 Korean women were diagnosed breast cancer. In Austria 4836 new cases were reported.

In the same year 1573 Korean women died of breast cancer and in Austria malign breast tumor caused 1583 deaths in females (Fig. 1).

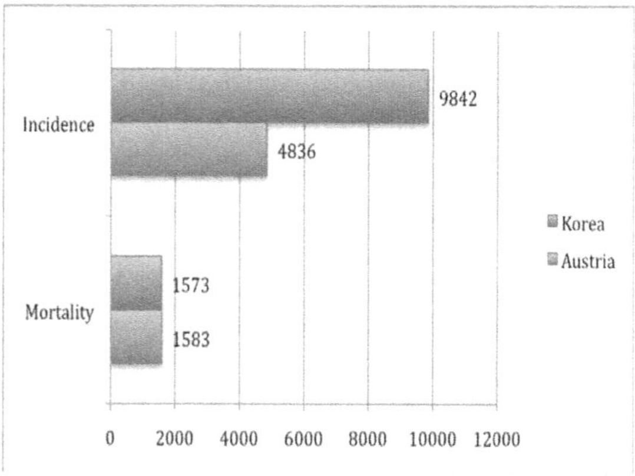

Fig. 1 Breast Cancer Incidence and Mortality: differences between Korea and Austria in 2005

A comparison of age-adjusted breast cancer incidence rate for Korean to Austrian demonstrated a much higher incidence of breast cancer in Austria (Fig. 2). The mortality is also higher in Austria.

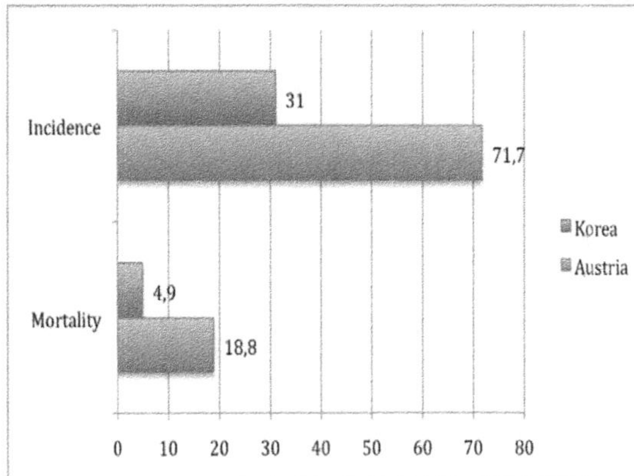

Fig. 2 Age-adjusted Incidence and Mortality Rates: differences between Korea and Austria in 2005

The incidence of breast cancer in Korea is still relatively low compared to Austria, but there is a rapid increase in incidence. Within 7 years (1999-2005) the age-adjusted incidence rate rose from 20,9 to 31 per 100.000 (Fig. 3).

A recent analysis shows that the increase of breast cancer in Korea is the highest in the world [6].

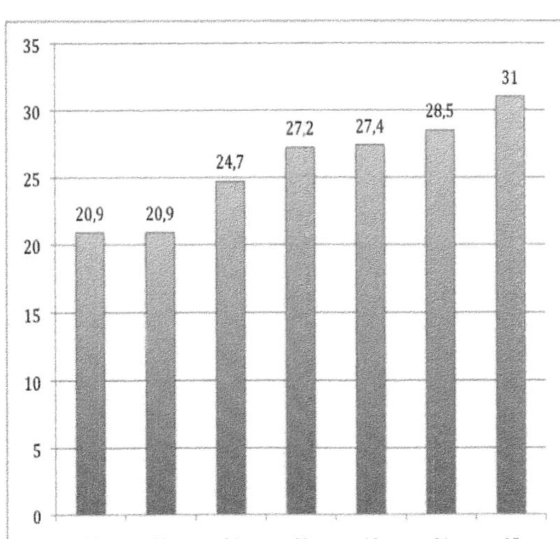

Fig. 3 Breast Cancer Incidence in Korea 1999-2005

The age-specific distribution of breast cancer in Koreas differs from that in Austria. In Western countries there is a linear increase of incidence with age. In Korea an inverted V-curve can be observed in age-specific incidence rates (Fig. 4).

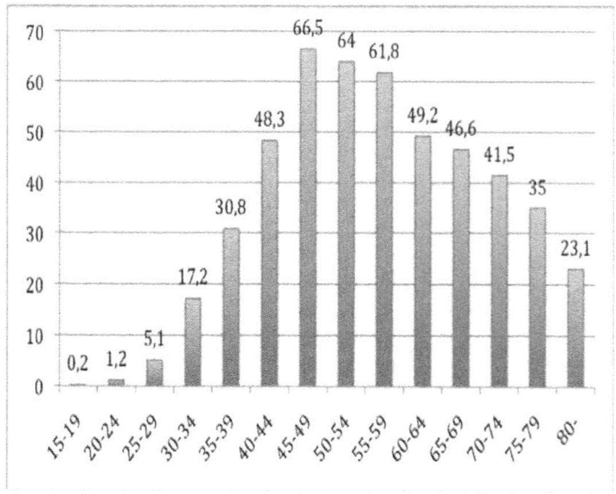

Fig. 4 Age-specific Breast Cancer Incidence in Seoul 1993-1997 Local cancer registry

The following graphic (Fig. 5) demonstrates the difference in pattern of age-specific incidence in Korea compared to Germany. The data was derived from Local cancer registry in Seoul and the Robert Koch Institut in Germany.

The highest incidence rate occurs in Korean women in their 40s.

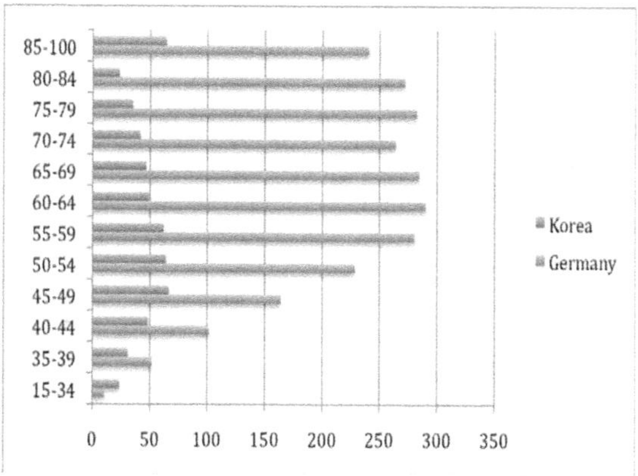

Fig. 5 Age-specific Breast Cancer Incidence: differences between Korea and Germany

Screening

The Korean government has already released a guideline of National Cancer Screening Program (NCSP). All women aged 40 or older are offered mammography and clinical breast exam every second year [8].

In Austria there is a similar screening offer of mammography and clinical examination every 1 to 1,5 years. Depending on the mammography result ultrasound is administered.

Breast Conserving Surgery

Modified radical mastectomy is the most frequently performed surgical method in Korea. Only about 25% of the breast cancer patients receive breast conserving surgery at the Asan Medical Center, in which about 750 cases of primary breast cancer is treated annually. This number represents 10% of all Korean breast cancer cases. However, breast-conserving surgery has increased continuously in Korea in the last decades [9].

In Austria certified breast care centers, such as Brustgesundheitszentrum Linz, have the aim to keep the breast-conserving rate about 80% and more [10,11].

Discussion

Study Group

The Austrian Breast and Colorectal Cancer Study Group (ABCSG) is one of the largest and successful study groups all over the world. Up to 40% of Austrian breast cancer patients participate in ABCSG trials. In Korea, a nationwide organization for prospective randomized multicenter trials for breast cancer has not been established yet. This might be a result of relatively low incidence of breast cancer in Korea, but since the incidence is rapidly increasing, a study group could support breast cancer therapy. Since 1996 clinical data about malign breast tumors have been collected by Korean Breast Cancer Society (KBCS). Based on this data further studies could be made.

Since the Korean breast cancer shows different clinical patterns, multicenter trials could support Korean public health.

Lifestyle

Evidence shows that the risk of breast cancer is higher for Korean Americans than for sedentary Koreans [12]. This evidence implicates that lifestyle changes of Koreans might play a role in development of breast cancer. The changes in socioeconomic status, food intake, weight of Koreans are notable. Recent studies also show a positive association between height and cancers of the breast [13]. The correlation between Western lifestyle and breast cancer need more investigations.

Surgical Procedures

The high rate of mastectomy (75%) in Korea might result from the lower average breast volume of Korean women. Specific explanations for this high mastectomy rate are unavailable.

Prevention Strategies

The best option to reduce breast cancer mortality is early detection/ diagnosis. The Korean government has already released a guideline for mammography screening [8]. Every screening program needs to be monitored to ensure quality standards.

An important issue is how many Korean women will follow the breast cancer screening program. Public health strategies need to be set to raise awareness in Korean women.

References

[1] Ferlay, J., et al.: GLOBOCAN 2000: Cancer Incidence, Mortality and Prevalence Worldwide, IARC Cancer Base No. 5 (1.0) (2001)

[2] Jung, K.W., et al.: Cancer Statistics in Korea: Incidence, Mortality and Survival in 2005. J. Korean Med. Sci. 24, 995–1003 (2009)

[3] Yoo, K.Y., et al.: Epidemiology of Breast Cancer in Korea: Occurrence, High-Risk Groups, and Prevention. J. Korean Med. Sci. 17, 1–6 (2002)

[4] Austrian Breast and Colorectal Cancer Study Group (2010), `http://www.abcsg.org/globalPlayer.html`

[5] Statistik Austria (2010), `http://www.statistik.at/web_de/statistiken/gesundheit/krebserkrankungen/index.html`

[6] Bray, F., et al.: The changing global patterns of female breast cancer incidence and mortality. Breast Cancer Res. 6, 229–239 (2004)

[7] Robert Koch Institut- Krebsregisterdaten (2002), `http://www.rki.de/cln_169/DE/Content/GBE/DachdokKrebs/krebs__node.html?__nnn=true`

[8] Ministry of Health and Welfare ROK. National Cancer Control Guideline (2004)

[9] Son, B.H., et al.: Changing Patterns in the Clinical Characteristics of Korean Patients With Breast Cancer During the Last 15 Years. Arch Surg.: 141, 155–160 (2006)

[10] Austrian Breast and Colorectal Cancer Study Group (2010), `http://www.abcsg.at/abcsg/html_patienten/vorteile.html#brusterhaltung`

[11] Brustgesundheitszentrum Linz (2010), `http://www.brust-gesundheitszentrum.at/index_html?sc=48524310`

[12] Lee, J., et al.: Cancer Incidence Among Korean-American Immigrants in the United States and Native Koreans in South Korea. Cancer Control 14(1), 79–85 (2007)

[13] Sung, J., et al.: Height and site-specific cancer risk: A cohort study of a korean adult population. Am J. Epidemiology 170(1), 53–64 (2009)

Overexpression and Unique Rearrangement of VH2 Transcripts in Immunoglobulin Variable Heavy Chain Genes in Ankylosing Spondylitis Patients

Yeon Joo Kim[1], Nayoung Kim[1], Min-Kyung Lee[2], Hyo-Jin Choi[2], Han Joo Baek[2,*], and Chang-Hoon Nam[1,*]

[1] Korea Institute of Science and Technology - Europe Forschungsges. mbH, Campus E71, D66123, Saarbrücken, Germany
chang@kist-europe.de

[2] Department of Rheumatology, Gachon University School of Medicine, Gil Hospital, Incheon, Republic of Korea
baekhj@gilhospital.com

Abstract. To identify immunoglobulin (Ig) variable heavy chain (VH) gene usages in Korean ankylosing spondylitis (AS) patients, expression level of VH2 genes from peripheral blood mononuclear cells (PBMCs) of 8 AS patients and 9 healthy donors was analysed by quantitative real-time PCR (Q-PCR). Q-PCR results demonstrated VH2 genes were overexpressed in AS patients (Relative amount of mRNA of VH2 genes to a house¬keeping gene, 7.13±7.77 vs, 0.68±0.55; $p<0.0001$). The sequence analysis revealed the majority of them contained CDC42 binding protein kinase beta (CDC42 BPB) genes. The insertion of CDC42 BPB gene was confirmed by PCR with primers corresponding CDC42 BPB and CH genes. Our study revealed VH2 overexpression and unique rearrangement in Ig VH genes from peripheral blood of AS patients. This may imply aberrant immunoglobulin gene rearrangement in B cell occurs in Korean AS patients, which requires further investigation.

1 Introduction

Ankylosing spondylitis (AS), a prototype of spondyloarthritis (SpA), is a chronic inflammatory arthritis that mainly affects the sarcroiliac joints and the spine. It is characterized by peripheral arthritis, enthesitis, and extraskeletal features, such as uveitis and inflammatory bowel disease. The pathogenesis of AS/SpA is not fully elucidated, but much has been learned regarding this condition in recent decades. Human leukocyte antigen (HLA)-B27, a major histocompatibility complex (MHC) class I allele, is a key genetic factor in AS/SpA (Brewerton, Hart et al., 1973; Schlosstein, Terasaki et al., 1973). Enteric bacteria may also play a role in causing the diseases as an environmental component (Taurog, Richardson et al., 1994; De Keyser, Elewaut et al., 1998). Populations of immune cells have been

* Corresponding author.

M.-W. Han and J. Lee (Eds.): EKC 2010, SPPHY 138, pp. 113–122.
springerlink.com

implicated in the pathogenesis of AS/SpA. T cells, particularly CD4+ T cells, are required for developing the disease in rats (Breban, Fernandez-Sueiro et al., 1996; May, Dorris et al., 2003). There has been evidence that macrophages participate in the pathogenesis of the disease (Baeten, Demetter et al., 2002; McGonagle, Marzo-Ortega et al., 2002). Recent data have demonstrated that Th17 cells, interleukin-17-producing effector T helper cells, contribute to AS/SpA inflammation (Jandus, Bioley et al., 2008; Shen, Goodall et al., 2009). However, specific autoantibodies have not yet been identified in patients with AS/SpA, and the role or relevance of B cells in the pathogenesis of the disease is unclear.

Analyses of immunoglobulin (Ig) variable region (V) gene usages have revealed differences in the basic Ig V repertoire of patients with B cell- and/or T cell-mediated autoimmune diseases, including systemic lupus erythematosus (Isenberg, Spellerberg et al., 1993; Stevenson, Longhurst et al., 1993), myasthenia gravis (Sims, Shiono et al., 2001), rheumatoid arthritis (Robbins, Kenny et al., 1990; Ermel, Kenny et al., 1997), and Sjögren syndrome (Dorner, Hansen et al., 2002), compared to healthy controls. Such differences could be derived from intrinsic abnormalities in the generation of Ig V genes or B cell development and functions (Foreman, Van de Water et al., 2007). Thus, the analysis of Ig V gene usages can offer new insights into possible pathogenic role of B cells in diseases.

A previous study of heavy chain variable segment (VH) germline gene usages in synovial B cells from AS patients demonstrated that the majority of rearranged Ig VH genes belonged to the VH3 genes (Voswinkel, Weisgerber et al., 2001). However, it did not include a direct comparison to healthy donors and some of the VH germline genes were overlooked due to the incomplete VH PCR primer sets. Therefore we investigated VH germline gene usages of patients with AS compared to healthy donors using PCR with additional primer sets.

2 Experimental Procedures

Subjects

Peripheral blood (PB) was collected from 9 healthy controls and from 8 patients with AS who visited the rheumatology clinic at Gachon University Gil Hospital. The patients with AS met the Modified New York Criteria (van der Linden, Valkenburg et al., 1984) and were taking non-steroidal anti-inflammatory drug and sulfasalazine regularly. Age, sex, disease duration, erythrocyte sedimentation rate (ESR), serum C reactive protein (CRP) and HLA-B27 positivity were assessed (Table 1). The validated Korean version of Bath AS Disease Activity Index (BASDAI) (Garrett, Jenkinson et al., 1994) for AS patients were calculated when their PB was collected. The study was approved by Local Research Ethical Committee of Gachon University Gil Hospital and written informed consent was obtained from all the patients and healthy volunteers.

Table 1 Demographic and clinical characteristics of subjects

	AS (n=8)	HC (n=9)
Age (years)*	38.6 ± 13.1	30.2 ± 2.3
Sex (M:F)	7:1	8:1
Disease duration (years)	11.4 ± 7.1	
History of uveitis	3 (37.5%)	
History of enthesitis	2 (25%)	
BASDAI*	4.2 ± 2.3	
BASFI*	2.6 ± 2.2	
ESR (mm/hr)*	37.0 ± 35.8	
CRP (mg/dL)*	1.7 ± 2.2	
HLA-B27 positivity	8 (100%)	

AS= ankylosing spondylitis, HC= healthy controls BASDAI= Bath Ankylosing Spondylitis Disease Activity Index BASFI= Bath Ankylosing Spondylitis Functional index
* mean ± SD

Quantitative real-time PCR (Q-PCR)

PB was collected into sterile, heparinized tubes. PB mononuclear cells (PBMCs) were isolated from whole blood using a Ficoll–Hypaque gradient. Total RNA was isolated from PBMCs by RNAeasy mini kit (Qiagen, Hilden, Germany) and cDNA was synthesized by Maxime™ RT PreMix (Oligo(dT)15 primer) Kit (Intron Biotechnology, Korea) following manufacturers' instructions. PCR and Q-PCR were performed with primers as shown in Table 2 (Cowell, Kim et al., 1999; Van Esch, Reparon-Schuijt et al., 2003; Dheda, Huggett et al., 2004). For thorough representation of VH germline genes, two novel VH forward primers, which we refer to as VH2*a and VH4*a forward primers for Q-PCR, were added to VH PCR primer sets used in previous studies (Voswinkel, Weisgerber et al., 2001). The sequences of VH2a and VH4a forward primers previously used were different from those for VH2*a and VH4*a, as shown in Table 1. Primers for human acidic ribosomal protein (HuPo) gene, as a house-keeping gene, were also used since it is known to be more reliable than β-actin gene as control for the Q-PCR in PBMC (Dheda, Huggett et al., 2004). PCR mixture without DNA was included in each experiment as negative control. The amplification condition was 95°C for 30 sec, 50-60°C for 1 min, and 72°C for 1 min for 30-40 cycles. Primers were produced by MWG (Ebersberg, Germany). Reagents for Q-PCR were purchased from Stratagene (La Jolla, USA). PCR products were analyzed by gel electrophoresis or by MxPro™ QPCR software (Stratagene, La Jolla, USA). The relative amount of transcripts of target genes compared to those of a housekeeping gene was calculated as follows;

$$\Delta Ct = Ct(\text{experimental}) - Ct(\text{housekeeping}),\ R = 2\text{-}(\Delta Ct).$$

Table 2 Oligonucleotide primers used to evaluate VH gene usage and confirm CDC42BPB inserted gene structure

Primer set for Q-PCR	
Human VH forward primers	
HuVH1a for	5' – CAG GTG CAG CTG GTG CAG TCT GG – 3'
HuVH2a for	5' – CAG GTG AAG TTA AGG GAG TCT GG – 3'
HuVH3a for	5' – GAG GTG CAG CTG GTG GAG TCT GG –3'
HuVH4a for	5' – CAG GTG CAG CTG CAG GAG TCG GG – 3'
HuVH5a for	5' – GAG GTG CAG CTG TTG CAG TCT GC – 3'
HuVH6a for	5' – CAG GTA CAG CTG CAG CAG TCA GG – 3'
HuVH2*a for	5' – CAG ATC ACC TTG AAG GAG TCT GG – 3'
HuVH4*a for	5' – CAG GTG CAG CTA CAG CAG TGG GG – 3'
Human JH reverse primers	
HuJH1-2 rev	5'– TGA GGA GAC GGT GAC CAG GGT GCC – 3'
HuJH3 rev	5' – TGA AGA GAC GGT GAC CAT TGT CCC – 3'
HuJH4-5 rev	5' – TGA GGA GAC GGT GAC CAG GGT TCC – 3'
HuJH6 rev	5' – TGA GGA GAC GGT GAC CGT GGT CCC – 3'
HuJH7 rev	5' – TGA CCG TGG TCC CTT GGC CCC AGA – 3'
HuPo(House-keeping gene)	
HuPo for	5'-CCA TTC TAT CAT CAA CGG GTA CAA-3'
HuPo rev	5'-AGC AAG TGG GAA GGT GTA ATC C-3'
Primer set for second PCR	
Human VH2*a forward primers with NcoI restriction sites	
HuVH2*a-Nco1	5' – AGC CGG CCA TGG CCG CAG ATC ACC TTG AAG GAG TCT GG – 3
Human JH reverse primers with XhoI restriction sites	
HuJH1-2-Xho1	5' –TCC ACC GCT CGA GAC TGA GGA GAC GGT GAC CAG GGT GCC–3'
HuJH3-Xho1	5'– TCC ACC GCT CGA GAC TGA AGA GAC GGT GAC CAT TGT CCC–3
HuJH4-5-Xho1	5'–TCC ACC GCT CGA GAC TGA GGA GAC GGT GAC CAG GGT TCC–3'
HuJH6-Xho1	5'– TCC ACC GCT CGA GAC TGA GGA GAC GGT GAC CGT GGT CCC–3'
HuJH7-Xho1	5'– TCC ACC GCT CGA GAC TGA CCG TGG TCC CTT GGC CCC AGA–3'
Primer set to confirm CDC42BPB insert	
HuCDC42 BPB Forward primer	
HuCDC42-FOR	5'-GAG CAC TGG CCA AGC ACT A-3'
Human Cμ, Cγ and Cεreverse primer	
Hu C μ rev	5'-TCC AGG AGA AAG TGA TGG AG-3'
Hu C γ rev	5'-GTC TTG GCA TTA TGC ACC TC-3'
Hu C ε rev	5'-CGG ATG GGC TCT GTG TGG-3'

Sequence analysis

The second PCR was performed with Q-PCR products from selected VH2* over-expressed patient samples (number 4, 6, 7, 8, and 9) as a template to introduce NcoI and XhoI site at 5' and 3' end of first PCR products, respectively. The PCR products were cloned into pIT2 vectors (Enever, Tomlinson et al., 2005) and 60-80 colonies were selected to confirm inserted genes by PCR screening using primer set for second PCR (Table 1). Approximately 20 samples of each group were analyzed by sequencing (MWG, Germany). The sequence results were analyzed by homology comparison with Ig Blast databases (http://www.ncbi.nlm.nih.gov/igblast) and IMGT databases (http://imgt.cines.fr).

To identify incorporation of CDC42 BPB intron fragments, PCR amplification was performed with primers for CDC42 BPB intron gene and human Ig constant region (C) genes (Table 1). Reaction conditions were initial denaturation at 95OC for 15 min, followed by 30 cycles of 94°C for 30 sec, 58°C for 30 sec and 72°C for 1 min, and then 72°C for 10 min for final extension. PCR products were analyzed by 2% agarose gel electrophoresis.

Statistics

A non-parametric statistic method, two tailed Mann-Whitney U test was performed to test the significance of differences in mRNA amount of individual VH gene families between AS patients and healthy volunteers (SPSS, Ver. 12.0. Chicago, IL). To test for correlations between the expression level of VH2*a genes and clinical variables, Spearman's rank correlation coefficient was used.

3 Results

Profile of subjects

The demographic and clinical characteristics of patients and controls are shown in Table 1. The mean age of AS patients was older than controls (38.6 and 30.2 years, respectively), but it was not statistically significant.

Overexpressed VH2* genes in AS patients

To analyze VH2 gene usages in detail, Q-PCR was performed with RNA from PBMC in AS patients. After we checked out that primer sets used in previous studies (Voswinkel, Weisgerber et al., 2001) can cover VH germline sequence from Ig Blast databases, we realized that 8 sequences of IGHV2 subfamilies (IGHV2-5*01, 5*04, 5*05, 5*06, 5*07, 5*10, 70*09 and 70*12) and 10 sequences of IGHV4 subfamilies (IGHV4-34*01, 34*03, 34*04, 34*05, 34*06, 34*07, 34*08, 34*11, 34*12 and 59*10) were not covered with those primer sets (Cowell, Kim et al., 1999; Van Esch, Reparon-Schuijt et al., 2003) (Table 2). Therefore additional primers were designed to cover complete VH germline gene and designated them as VH2*a and VH4*a.

The results from reverse transcriptase PCR with individual samples and from Q-PCR with pooled samples demonstrated that there were no significant differences in PCR profile between AS patients and control group, that was produced by PCR primer sets for VH1a-VH6a and for VH4*a (Fig. 1A and B). The results confirmed the previous study showing VH3a was dominantly expressed (Voswinkel, Weisgerber et al., 2001). Interestingly, VH2* genes represented by VH2*a forward primer were overexpressed exclusively in AS patients (patient sample numbers 4, 5, 6, 7, and 8 as shown in Fig. 1A). Thus, Q-PCR with individual samples was performed to quantify the amount of VH2* transcripts. Significant difference was shown in the level of expression of VH2* genes between healthy donors and AS patients (relative amount of mRNA of VH2* genes to human acidic ribosomal protein (HuPo),

0.68±0.55 [mean±SD] and 7.13±7.77, respectively; $p<0.0001$, Fig 1C). There were no correlations of the expression level of VH2* with clinical variables such as sex, age, disease duration and inflammatory parameters (data not shown).

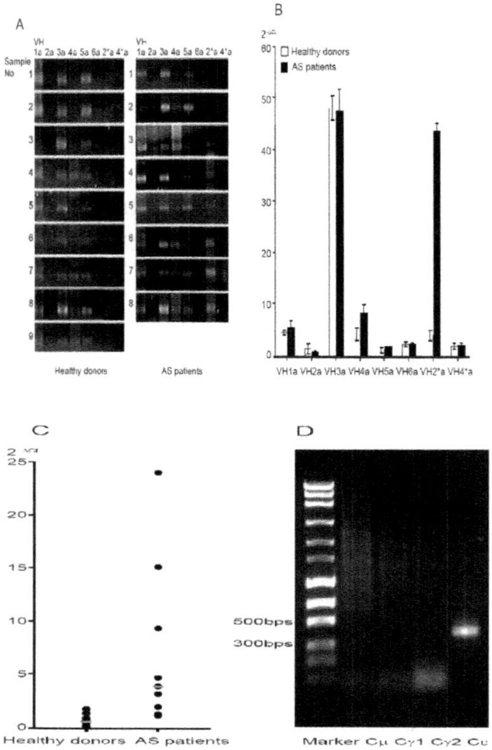

Fig. 1 VH2 genes were overexpressed in AS patients. (A) Comparison of VH gene usages in PBMCs from healthy donors and AS patients. Reverse transcriptase PCR was performed with RNA from PBMCs of healthy donors and AS patients using primers shown in Table 2. VH2* genes were overexpressed exclusively in AS patients (patient sample number 4, 5, 6, 7 and 8). N= 9 of healthy donors and 8 of AS patients. **(B)** The relative expression levels of patients and healthy donors' each VH gene. QPCR was performed with pooled cDNA from PBMC of healthy donors and AS patients using primers specific for VH1a-6a genes. The relative amount of transcripts of target genes compared to those of a housekeeping gene was calculated as follows; $\Delta Ct=Ct(experimental)-Ct(housekeeping)$, $R=2^{-(\Delta Ct)}$. The data shows the mean and standard deviation of triplicate PCR amplications **(C)** Q-PCR was performed with individual cDNA from each PBMC of healthy donor and AS patient using primers specific for VH2* genes. The relative amount of mRNA of VH2* genes to HuPo was calculated as above. The relative amount of mRNA of VH2* genes to HuPo in AS patients was significantly higher compared to healthy donors ($P<0.0001$). N= 9 of healthy donors and 8 of AS patients. **(D)** CDC42 BPB intron sequences were inserted in Ig genes in AS patients. PCR was performed with cDNA from PBMCs of healthy donors and AS patients using primers specific for CDC42 BPB genes and Ig C epsilon genes.

Gene structure of VH2* PCR products

The PCR products from samples demonstrating high expression level of VH2* genes were cloned and sequenced. Sequence analysis revealed a short fragment from CDC42 BPB genes incorporated into major part of cloned VH2* PCR products from patient sample number 4, 5, 6, 7, and 8. This is an intron fragment located in the region encompassing 252 bps (36096-36348) of 125-kb of CDC42 BPB which maps to 14q32.32 (Fig 2) (Moncrieff, Bailey et al., 1999) and is assumed to be incorporated in-between a short stretch of VH2 and DH6.

Another PCR using primers corresponding to sequences in the middle of the intron of CDC42 BPB gene and Ig constant regions (Cμ, Cγ and Cε) was set up to confirm that the intron fragment from CDC42 BPB genes was indeed incorporated into Ig heavy chain gene segments (Fig 1D). The results demonstrated the expected band size, about 330 bps, which appeared exclusively by a Cε primer. Therefore these results imply that an intron fragment of the CDC42 BPB gene, suggested by sequence analysis, was incorporated in between VH2 and DH6-JH3-Cε of rearranged Ig genes (Fig 1D).

4 Discussion

Immunoglobulins consist of the light (L) and heavy (H) chains, each of which has variable and constant regions. The human VH segments are located in three loci; chromosome 14, 15 and 16 but only the chromosome 14 locus contains the JH segments that are essential for somatic generation of the VH genes.

As for VH germline gene usages in AS patients, a previous report demonstrated over-representation of VH5 and under-representation of VH4 from the AS synovial B lymphocytes compared with the germline representation (Voswinkel, Weisgerber et al., 2001). In this study, we used a novel primer set including VH2*a, representing some of the IGHV2 genes in the VH2 germline gene family, which had been absent in the previous study. VH2* genes were overexpressed only in AS patients and the level of expression was significantly higher in PB of AS patients compared to those of healthy donors. It suggests that restricted germline gene family may be selected in PB of AS patients.

The Ig V gene repertoire in PB was not compared with either synovial tissues or synovial fluid in the same patients in this study. The Ig V gene repertoire in PB may be different from that in inflammatory joints. PB contains a population of recirculating memory B cells that have encountered a wide diversity of antigens over the patient's lifetime, whereas inflammatory joint tissues may have the subset of B cells responding to antigen and undergoing antigen-driven response. Thus comparing Ig contents of synovial B cells with those of PB should be interesting.

Strikingly the sequence analysis and homology search of overexpressed VH2*PCR products revealed unexpected features of VH gene structure. In the sequences of VH2* PCR products, a short stretch of CDC42 BPB intron gene was found (Fig. 2). It is worth noting that this intron segment is located in chromosome 14q where human Ig VH gene locus is present (Cook, Tomlinson et al., 1994). Ig gene rearrangement requires recombination signal sequences (RSS) that consist of

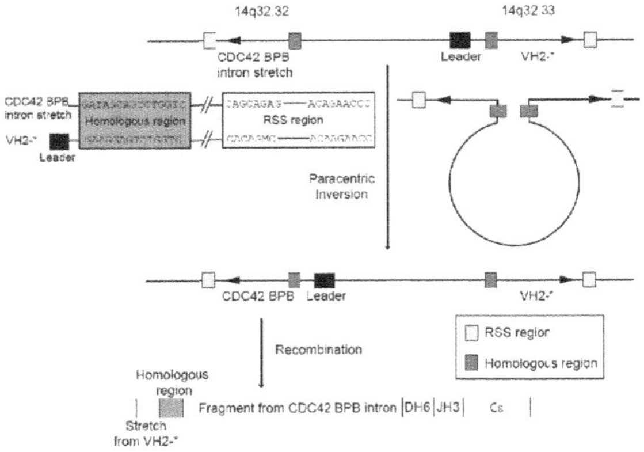

Fig. 2 Proposed gene structure of rearranged VH2* genes in AS patients. CDC42 BPB intron fragment could be paracentrically inverted into VH2 genes. Both genes are located in chromosome 14q32. The sequence homology search revealed possible RSS sequences close to the inserted CDC42 BPB intron fragment. The resulted rearranged Ig gene has a part of the VH2 genes, CDC42 BPB intron sequences, DH6, JH3, and Cε in order.

a heptamer, 23 nucleotides (nts), and a nonamer in this order (Jung, Giallourakis et al., 2006). A short stretch of intron sequences of CDC42 BPB was found between VH and DH genes. The 5' end of this fragment includes homologous region (15 nts) to that of the VH2 germline gene (Fig 2). Interestingly, further analysis of CDC42 BPB genomic sequences identified a nonamer of RSS, which is located following an insertion site of this fragment by 18 nts, exists in CDC42 BPB genes. In addition, a heptamer-like sequence (CAGCAGAG), which has one G base added feature to generally known heptamer of RSS (CACAGAG), also appears in the end of the inserted CDC42 BPB intron genes, as seen in Fig 2. This gene structure implies that unusual gene rearrangement or recombination may occur during construction of VH gene segments during B cell development in some of the AS patients. Paracentric inversion following recombination activating gene (RAG)-based recombination might be suggested to explain unique rearranged VH2* gene structure (Fig 2). However, it should be mentioned that this hypothesis is not perfectly fitted with the norm of RAG-based recombination as the distance between heptamer and nonamer in CDC42 BPB gene is shorter and heptamer¬like sequence in CDC42 BPB needs to have one base deletion to become proper heptamer for recombination. The questions how these unfitted features could be modified in some AS patients and whether the unique rearrangements were found only in Korean AS patients are still remaining. Further study is required to determine these questions.

To date, evidence that B cells are involved in AS/SpA pathogenesis has been relatively scant. Nevertheless, B cells and plasma cells are consistently present in inflammatory lesions (McGonagle, Marzo-Ortega et al., 2002; Appel, Kuhne et al., 2006). In a recent open label phase-II clinical trial, rituximab, a chimeric monoclonal antibody against CD20 on the surface of B cells, showed significant efficacy in TNF-naïve patients with AS (Song, I., et al., 2009). Our study, along with Voswinkel's work (Voswinkel, Weisgerber et al., 2001), showed abnormal molecular event in B cells from AS patients. Taken together, these suggest that B cells may play a role in developing AS. Since the autoantibodies are not identified in AS, B cells might contribute to AS/SpA pathogenesis by (auto)antigen presentation to T cells or cytokine release rather than by antibody production. In a mouse model of proteoglycan-induced arthritis, B cells contributed to the disease by presentation of autoantigen and subsequent T cell activation as well as by producing specific antibodies (O'Neill, Shlomchik et al., 2005). Additional sequence analysis of rearranged VH2* gene structure is currently being performed, and it is planned to screen binding partners from serum of AS patient against this aberrant gene product. It would be also informative to determine whether the rearranged Ig actually produces protein. However, it is possible that the aberrantly rearranged Ig genes are not functional, thus no protein could be produced. Aberrant Ig transcripts could still affect other Ig protein production or composition in a way or another. Further investigations are required to verify that distinct rearrangement of Ig VH genes are involved in B cell development or function in terms of AS pathogenesis.

In conclusion, this study revealed VH2 overexpression and unique rearrangement in Ig VH genes from PB of AS patients. This may imply aberrant immunoglobulin gene rearrangement in B cell occurs in Korean AS patients, which awaits further investigation.

References

[1] Appel, H., et al.: Immunohistologic analysis of zygapophyseal joints in patients with ankylosing spondylitis. Arthritis Rheum (2006)

[2] Baeten, D., et al.: Macrophages expressing the scavenger receptor CD163: a link between immune alterations of the gut and synovial inflammation in spondyloarthropathy. J. Pathol. (2002)

[3] Breban, M., et al.: T cells, but not thymic exposure to HLA-B27, are required for the inflammatory disease of HLA-B27 transgenic rats. J. Immunol. (1996)

[4] Brewerton, D.A., et al.: Ankylosing spondylitis and HL-A 27. Lancet (1973)

[5] Cook, G.P., et al.: A map of the human immunoglobulin VH locus completed by analysis of the telomeric region of chromosome 14q. Nat. Genet. (1994)

[6] Cowell, L.G., et al.: Enhanced evolvability in immunoglobulin V genes under somatic hypermutation. J. Mol. Evol. (1999)

[7] De Keyser, F., et al.: Bowel inflammation and the spondyloarthropathies. Rheum Dis. Clin. North Am. (1998)

[8] Dheda, K., et al.: Validation of housekeeping genes for normalizing RNA expression in real-time PCR. Biotechniques (2004)

[9] Dorner, T., et al.: Immunglobulin repertoire analysis provides new insights into the immunopathogenesis of Sjogren's syndrome. Autoimmun. Rev. (2002)

[10] Enever, C., et al.: Engineering high affinity superantigens by phage display. J. Mol. Biol. (2005)

[11] Ermel, R.W., et al.: Analysis of the molecular basis of synovial rheumatoid factors in rheumatoid arthritis. Clin. Immunol. Immunopathol. (1997)

[12] Foreman, A.L., et al.: B cells in autoimmune diseases: insights from analyses of immunoglobulin variable (Ig V) gene usage. Autoimmun. Rev. (2007)
[13] Garrett, S., et al.: A new approach to defining disease status in ankylosing spondylitis: the Bath Ankylosing Spondylitis Disease Activity Index. J. Rheumatol. (1994)
[14] Isenberg, D., et al.: Identification of the 9G4 idiotope in systemic lupus erythematosus. Br. J. Rheumatol. (1993)
[15] Jandus, C., et al.: Increased numbers of circulating polyfunctional Th17 memory cells in patients with seronegative spondylarthritides. Arthritis Rheum. (2008)
[16] Jung, D., et al.: Mechanism and control of V(D)J recombination at the immunoglobulin heavy chain locus. Annu. Rev. Immunol. (2006)
[17] May, E., et al.: CD8 alpha beta T cells are not essential to the pathogenesis of arthritis or colitis in HLA-B27 transgenic rats. J. Immunol. (2003)
[18] McGonagle, D., et al.: Histological assessment of the early enthesitis lesion in spondyloarthropathy. Ann. Rheum Dis. (2002)
[19] Moncrieff, C.L., et al.: Cloning and chromosomal localization of human Cdc42-binding protein kinase beta. Genomics (1999)
[20] O'Neill, S.K., et al.: Antigen-specific B cells are required as APCs and autoantibody-producing cells for induction of severe autoimmune arthritis. J. Immunol. (2005)
[21] Robbins, D.L., et al.: Serologic and molecular characterization of a human monoclonal rheumatoid factor derived from rheumatoid synovial cells. Arthritis Rheum. (1990)
[22] Schlosstein, L., et al.: High association of an HL-A antigen, W27, with ankylosing spondylitis. N. Engl. J. Med. (1973)
[23] Shen, H., et al.: Frequency and phenotype of peripheral blood Th17 cells in ankylosing spondylitis and rheumatoid arthritis. Arthritis Rheum (2009)
[24] Sims, G.P., et al.: Somatic hypermutation and selection of B cells in thymic germinal centers responding to acetylcholine receptor in myasthenia gravis. J. Immunol. (2001)
[25] Stevenson, F.K., et al.: Utilization of the VH4-21 gene segment by anti-DNA antibodies from patients with systemic lupus erythematosus. J. Autoimmun. (1993)
[26] Taurog, J.D., et al.: The germfree state prevents development of gut and joint inflammatory disease in HLA-B27 transgenic rats. J. Exp. Med. (1994)
[27] van der Linden, S., et al.: Evaluation of diagnostic criteria for ankylosing spondylitis. A proposal for modification of the New York criteria. Arthritis Rheum (1984)
[28] Van Esch, W.J., et al.: Human IgG Fc-binding phage antibodies constructed from synovial fluid CD38+ B cells of patients with rheumatoid arthritis show the imprints of an antigen-dependent process of somatic hypermutation and clonal selection. Clin. Exp. Immunol. (2003)
[29] Voswinkel, J., et al.: B lymphocyte involvement in ankylosing spondylitis: the heavy chain variable segment gene repertoire of B lymphocytes from germinal center-like foci in the synovial membrane indicates antigen selection. Arthritis Res. (2001)

Contribution of Time-Resolved Absorption Spectroscopy to Study Biological Questions

Byung-Kuk Yoo, Isabelle Lamarre, Jean-Louis Martin, and Michel Negrerie*

Laboratoire d'Optique et Biosciences, INSERM, Ecole Polytechnique,
91128 Palaiseau Cedex, France
michel.negrerie@polytechnique.fr

Abstract. In this report, we illustrate through the study of two allosteric heme proteins the contribution of time-resolved absorption spectroscopy to the understanding of fundamental biological mechanisms. The first studied protein is the endogenous nitric oxide receptor (guanylate cyclase, sGC) whose activation and deactivation mechanisms are not yet fully resolved. We show that the rebinding of the proximal histidine occurs in ~100 picoseconds in sGC, which is the very first step of its deactivation following NO release. We also show that synergistic action of CO together with an allosteric activator induces the cleavage of the bond between heme iron and proximal histidine. The second one is the prototype of allosteric protein, the dioxygen transporter hemoglobin (Hb). In Hb, we show that the motion of the iron atom, central to the heme, moves in ~18 picoseconds after NO binding; this motion represents the very first step of the allosteric T $\rightarrow$ R transition.

1 Introduction

The interaction of a diatomic (NO, O_2, CO) with a heme protein is determined by its chemical reactivity [1] but also by the protein structure and dynamics. For example nitric oxide (NO) dynamics after release from its heme binding site span several orders of magnitude in time [2, 3]. This diatomic ligand encounters several structural and energy barriers on its way, diffusing to or from the heme and its surroundings. All the elementary steps corresponding to the overcoming of these barriers influence the values of kinetics constants k_{on} and k_{off} (thus the affinity K_D). By analyzing the individual kinetic steps we can understand the mechanisms and access to the structural features determining the properties of the heme proteins. Many questions on heme proteins mechanisms can be directly addressed by time-resolved spectroscopies: 1) how do proteins control the 6-c to 5-coordinate transition? 2) what is the molecular mechanism by which the signal is transduced from the heme domain to the catalytic domain? 3) how does the NO-receptor get deactivated?

Time-resolved spectroscopies allow us to know all the transient steps from the release of NO from heme to the unliganded relaxed protein. This is not only a

* Corresponding author.

M.-W. Han and J. Lee (Eds.): EKC 2010, SPPHY 138, pp. 123–134.
springerlink.com

problem of measuring kinetic traces, but of identifying all the transient species involved during the process. This is not trivial if we consider that for example 97% of photodissociated NO geminately rebind to the heme of sGC in 7.5 ps [3] and that the 3% remaining population represent the deactivation of sGC and must be identified. Thus, it is crucial to determine the structural events and the associated energy barriers following ligand release and preceding ligand binding. The time scale of the diatomic ligand motion within the protein core is in the picosecond to microsecond range. However, if the starting system is an aqueous solution with a heme-protein and a diatomic (O_2, CO or NO, but also H_2S depending on the role of the protein), it is impossible to record a binding step faster than ~0.1 μs because the diffusion of interacting molecules in solution occurs much slower than the intramolecular structural events to be probed. To overcome this limitation, one must start from the well-defined system consisting of a heme protein liganded with a diatomic to its heme.

The population of bound ligand is in permanent exchange with the solution and the probability of thermal release for a particular diatomic ligand depends upon the protein structure and properties. Thus, probing the ligand motion and structural associated events in the ps-μs time-range is not possible in a thermally equilibrated system. Photodissociating the diatomic ligand with an ultra-short laser pulse (~0.1 ps) allows to set to 1 the probability of ligand release for a population of the bound protein. Then, the ligand can either rebind "immediately" (*i.e.* in less than ~20 ps) or migrate within the protein core and finally in solution (from where it may rebind again in a longer time). We are interested in identifying the structural events and processes, and measuring their kinetics and associated energy barriers.

With this aim, the bond between the iron and the diatomic is broken by the absorption of a light pulse (duration of 30 fs to 6 ns) and the absorption spectrum of the heme is then recorded by a second short pulse of same duration but with broad spectral band. The time delay between the photodissociating pulse and the probing pulse is then varied and numerous transient spectra can be recorded over a time range which depends upon the structural events under study. From 0.1 ps to 5 ns, this delay is optically varied (by moving a set of mirrors, playing with the speed of light) whereas the 5 ns to 1 s time-range is accessible only by electronic delay for firing the exciting laser with respect to the probing laser [4]. In this report, we describe studies based on time-resolved absorption, but the same principles for time-resolved spectroscopy are applied for infra-red [5], circular dichroism [6] as well as Raman spectroscopy [7, 8]. To illustrate the importance of transient spectroscopy in biology, we describe two studies on heme-proteins: the first one is the endogenous NO receptor in mammals, the soluble guanylate cyclase (sGC) and the second one is the O_2 transporter in blood cells, the hemoglobin. Both of them are allosteric proteins, with the ability of influencing the properties a binding site by ligating a diatomic to a remote site located in another subunit of the protein.

2 Experimental Procedures

Time-resolved Spectroscopy. Transient absorption measurements were performed with the pump-probe laser system previously described [9]. Basic idea of time-resolved pump-probe experiment is that it catches the change between two different states. A powerful laser light source (hereafter called "pump-probe") interacts with the sample (in our case, protein) and excites it into a non-equilibrium state. Right after the sample experiences a change toward a new unknown equilibrium. Reaching this new equilibrium may take place in tens of picoseconds as well as in the ns-μs time-range. This change can be monitored at different time delays and differentiated by sending a low-energy second laser light source which is called as a "probe pulse" into the sample. This second probe pulse is functioning like a camera and this pulse should be harmless to the sample not to damage it. The source of ultra-short pulse (<100 fs) is a dye laser pumped by a continuous wave Argon laser (~2.5 W) providing pulses at 620 nm at a repetition rate of ~ 95 MHz. These pulses are amplified by a YAG laser at 30 Hz, and the amplified pulse train is split in two equal beams used respectively for pumping and probing the sample. The splitting of the pulse train is thus the origin of the time delay in such time-resolved experiment. After this splitting, each pulse train is separately modulated to obtain precisely shaped pulses (energy, wavelength, duration, and spectrum).

In a typical pump-probe experiment, the delayed probe pulse is applied to detect the optical difference ΔA induced by the pump pulse (Fig. 1). Measuring the evolution of ΔA at different time delays allow us to detect structural changes. The photodissociation of NO was achieved with an excitation pulse at 564 nm whose duration was ~40 fs with a repetition rate of 30 Hz. White light continuum for obtaining a broad band probe pulse was created after passing 8-mm cell that contained H_2O (Soret) or D_2O (Band III). Both beams were focused to a spot of ~50 μm and spatially overlapped in the sample cell, which was continuously rotated perpendicular to the beams to ensure sample renewal between the shots. The transient absorption spectra after a variable delay between pump and probe pulses were recorded by means of a CCD detector. Up to 160 scans were recorded and averaged with a dwell time of 2 s at each point.

Brief scheme of the principle of the setup is displayed at the upper panel in Fig. 1 with experimental 3-D transient spectra (down) as an example. Since we record a series of transient spectra (375-475 nm for Soret) at different time delays, the result consists of a matrix of differential absorption as a function of time and wavelength. The global analysis of the data at Soret was performed by singular value decomposition (SVD) of the time-wavelength matrix such that all spectral components were identified in the multiple time windows ranging from 25 ps to 500 ps [10]. Thus, the kinetics presented here represents the evolution of the associated differential spectra, not those at a single wavelength. Alternatively, kinetics at a particular wavelength can be performed, provided the absorption band is identified (Fig. 1). Singular Value Decomposition (SVD) of the time-wavelength matrix could then be performed with a program calculating the decay associated spectra for each individual exponential and their amplitude. SVD analysis procedure allowed to separate the spectral components and their kinetic evolution within the data matrix. This is important in order to separate populations of proteins which behave differently and to separate the elementary structural steps of relaxation after photodissociation of the ligand. The kinetic components were fitted to a multi-exponential function, where the minimal number i of exponents was determined iteratively.

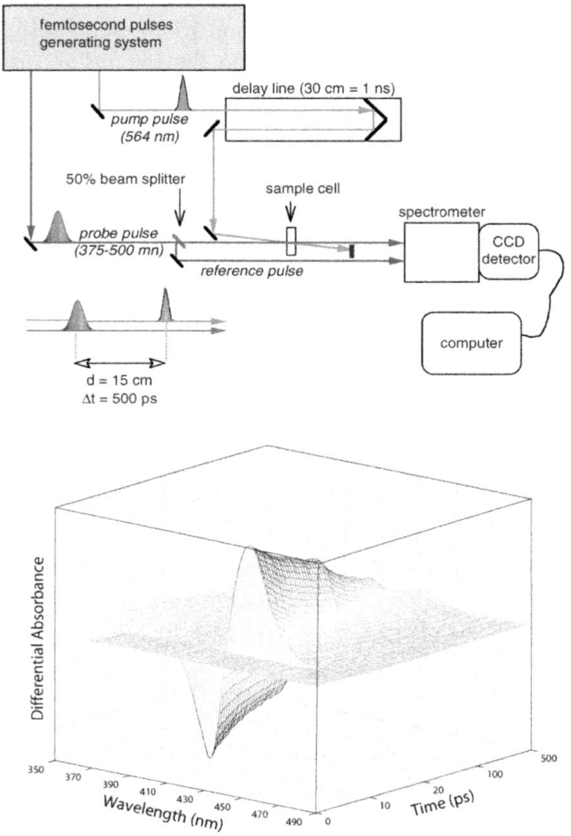

Fig. 1 Brief description of the transient absorption setup (up) and 3D transient absorption spectra of myoglobin with NO ligand (down). The dissociation of a diatomic or triatomic ligands (NO, CO, O_2, OH^-, H_2S) is performed by a first laser pulse. Structural changes in the protein are then probed by a delayed second laser pulse.

Preparation of Samples for Spectroscopy. sGC was purified from bovine lung using a previously described method [9] and human hemoglobin type A was purchased from Sigma. The samples were put in a 1-mm optical path length quartz cell, degassed. The sGC (70 µL, 40 µM) is already purified in the reduced state but Mb (100 µL, 5 mM) needs to be reduced by the sodium dithionite (4 mM). For preparing NO-liganded proteins, gas phase NO diluted in N_2 was directly introduced into the spectroscopic cell (~ 20 µM for 1 % NO in the aqueous phase) and wait for 10 min for equilibration between the gas phase and the solution. Equilibrium spectra were recorded at each step for monitoring the evolution of ligation. The absorbance of the sample was in the range 0.7-1 at the Soret maximum for 1 mm path length. For band III of hemoglobin, the absorbance was adjusted to 2.1 at 564 nm that is used to photoexcite the sample. Equilibrium spectra were recorded at each step to verify the stability of the sample.

3 Guanylate Cyclase Regulation

sGC is the endogenous NO receptor in mammals and it is our main subject protein. The rapidly diffusible messenger NO and the activation of sGC are involved in several physiological processes, including vascular pressure regulation, lung airway relaxation, and neural communication [11]. The functioning of the endogenous NO-receptor sGC is crucial in several diseases which comprises (among other) cardiovascular [12], pulmonary and renal pathologies and sGC is therefore a potential pharmacological target of great interest for cardiovascular treatment. Nitric oxide signaling pathway is also involved in apoptosis [13] and in tumor progression [14, 15]. Therefore, sGC inhibition is of great interest for studying and fighting tumor progression. A small number of sGC effectors exist so far, including some artificial compounds [16] and improving them or discovering new compounds requires the deep knowledge of sGC regulation at molecular level.

sGC catalyzes the formation of the second messenger cGMP from GTP upon NO binding. This heterodimeric protein possesses two subunits: the catalytic α-subunit harboring the GTP binding site and the regulatory β-subunit which contains a heme prosthetic group necessary for NO stimulation [11]. When NO binds to sGC, the covalent bond between the heme iron and the proximal histidine is broken and a 5-coordinate complex with NO is formed. This internal molecular event is assumed to trigger a structural change within the protein which induces the increase of catalytic activity. The entire mechanism of sGC functioning from cleavage of His-Fe bond to formation of cGMP is unknown. Since the sensing heme part is harbored on a different subunit than the site of GTP conversion, we assume that there is a "cross-talk" between both. This means that a structural change is necessary for this allosteric regulation. In order to have a molecular basis for developing a potential drug toward the sGC, it is essential to understand the activation and deactivation mechanisms.

Deactivation of guanylate cyclase. The aim of this experiment was to detect any structural relaxation after NO release from the heme iron. Since NO rebinds in 7.5 ps for 97 % of the population [3], we must analyze the remaining 3 % population for which NO does not rebind. The enzyme sGC was purified in the reduced form and reveals a characteristic Soret absorption maximum at 431 nm. After adding NO into the protein, the Soret band shifts to 399 nm on formation of the 5c-NO species [9].

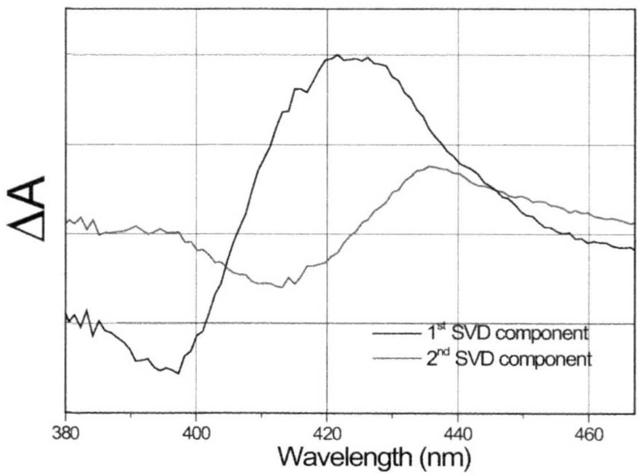

Fig. 2 Two spectral components obtained from SVD analysis for sGC-NO with YC-1. Geminate rebinding of NO component (SVD 1) within sGC heme pocket is clearly separated from the histidine rebinding component (SVD 2)

After measuring the transient absorption, we performed an SVD analysis, which resulted in two spectral components as shown in Fig. 2. The main component (SVD 1, black) corresponds to geminate rebinding of NO as shown by the decrease of the bleaching at 395 nm which is due to NO rebinding to the 4-coordinate heme. The SVD 2 spectral component whose intensity is lower appears with different maximum and minimum. The second component represents the contribution of His rebinding to sGC since induced absorption is located at 435 nm simultaneously with the decrease of the 4-coordinate heme spectrum. The time constant of the associated kinetics is ~100 ps, which represents the mean time needed for His rebinding to the sGC heme after NO release from the heme pocket. The picture is the following: once NO left the heme binding site, the histidine rebinds, leading to a structural change towards the resting-state. Thus, thermal equilibrium is sufficient to account for deactivation of sGC, whatever the fate of NO. The detection and identification of the proximal histidine recombination allowed us to record the first step of the deactivation mechanism of sGC.

Activation of guanylate cyclase Similarly to NO, CO is able to bind to the sGC heme and weakly activate the enzyme [17]. The binding of CO induces the formation of 6-coordinate sGC-CO complex and 4-fold increase in the rate of cGMP production. However, a benzylindazole derivative that activate sGC without participating to the bond with heme called YC-1 and its chemically similar product BAY 41-2272 from Bayer pharmaceutical company are known to show synergistic activation with CO. These two compounds constitute a novel class of sGC activators.

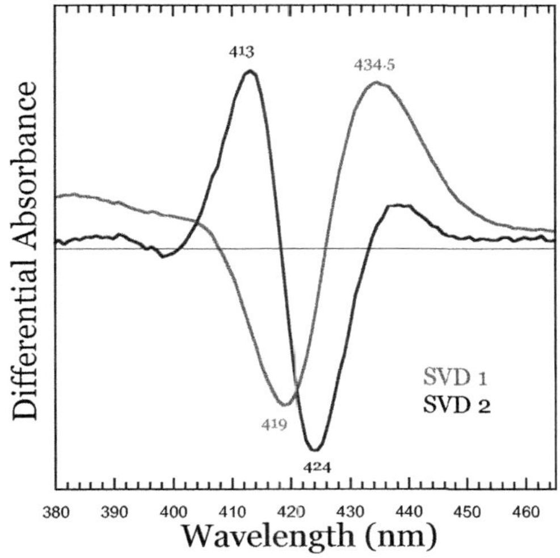

Fig. 3 Two spectral components obtained from SVD analysis for sGC-CO with BAY 41-2272. Geminate rebinding of CO component to 6c-CO (SVD 1) is clearly separated from that to 5c-CO (SVD 2).

After binding of the BAY 41-2272 compound into the sGC in the presence of CO and found two distinct species. Two different spectral SVD components are shown in Fig. 3. Two different bleaching minima and induced absorption peaks at different wavelengths represent each contribution of CO rebinding to different species. Induced absorption at 435.5 nm can be assigned to the formation of 5c-His heme species but that at 413 nm is produced by the formation of 4c-heme. Simultaneously, the bleaching at 419 nm can be attributed to the 6c-CO-heme but that at 424 nm to the 5c-heme-CO respectively. SVD 1 component represents the geminate recombination of CO to the 5c-heme iron and has slower decay time as shown in Fig. 4. In contrast, SVD 2 component represents CO recombination to the 4c-heme and is much faster than the CO rebinding to the 5c-heme. The bound effectors (YC-1 or BAY 41-2272) can alter the population of 6-coordinate sGC-CO species and produce a population of 5-coordinate sGC complex. Raman spectroscopy also shown a weak 5c-CO-heme in addition to the 6c-CO-heme produced by the effectors associated with an increased activity [18]. The faster CO rebinding is due to CO photodissociated from 5c-CO species, which rebind the very reactive 4c-heme.

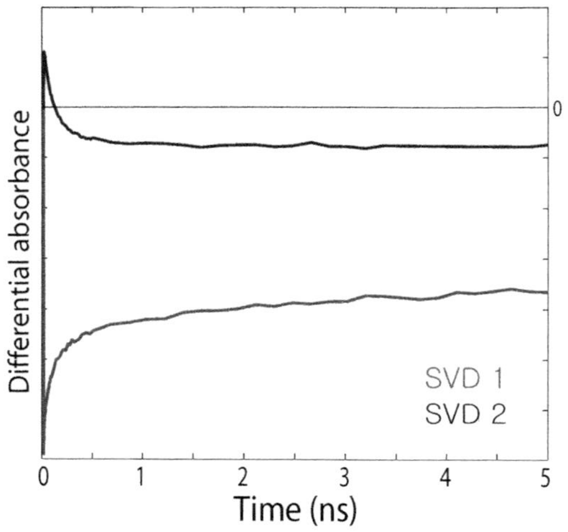

Fig. 4 Two kinetic components associated with the transient spectra of Fig. 3

The binding of YC1 or BAY 41-2272 effectors to sGC facilitates the breaking of the iron-histidine bond and increases the proportion of a 5-coordinate heme liganded with CO, and therefore increases the activated GC like NO activation. This is summarized in the Figure 5 which shows two different activation of sGC

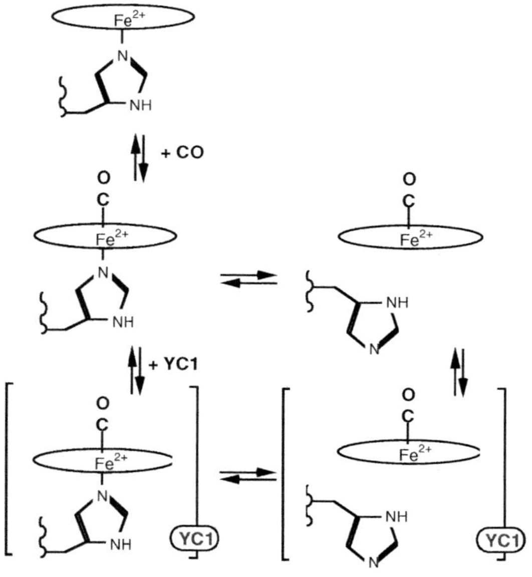

Fig. 5 Model for the synergistic activation of sGC induced by CO and YC-1. Two populations exist in the presence of CO (6c and 5c-heme). The equilibrium is shifted toward 5c-CO in the presence of YC-1 or BAY 41-2272.

by CO. 6c-heme and 5c-heme can exist separately and the induced formation of 5c-heme by YC-1 or BAY 41-2272 is shown with the distorted bracket which represents a structurally different sGC. Therefore, time-resolved spectroscopy allowed to explain the increased activity by the formation of 5c-CO species under the presence of the allosteric activators.

4 First Step of Allosteric Transition in Hemoglobin

As the very first events of activation and deactivation mechanisms, the binding and release of diatomics (O_2, NO or CO) trigger changes in protein structure. Thus, these local chemical events must be transmitted to the entire protein structure. This necessarily takes place through the heme iron motion. For example, this motion occurs from the proximal "out-of-plane" to the distal "out-of-plane" position for sGC activation, while this motion occurs from out-of plane to in-plane position of the iron with respect to the heme in the case of O_2 and NO binding to the O_2-carriers myoglobin and hemoglobin (5- to 6-coordinate iron). The out-of plane iron motion induced by NO or CO dissociation occurs faster than 0.3 ps as measured by transient absorption [19] or by Raman spectroscopy [7] and is thus considered quasi-instantaneous.

However, since the iron motion is coupled to protein backbone dynamics, the reverse in-plane motion induced by ligand binding is not necessarily instantaneous and has actually never been measured.

A theoretical study [20] focused on the absorption band III in 5-coordinate ferrous heme, assigned to a charge transfer transition from π-porphyrin molecular orbital to a d_{yz} iron orbital. This study concluded that intensity of band III is sensitive to the iron-heme plane distance, but rebinding of NO to heme proteins was never investigated by probing band III absorption kinetics.

We aimed at measuring the band III intensity kinetics in the picosecond time-range in order to compare it with Soret intensity, which reflects only the evolution of heme iron coordination, whereas band III reflects the iron position with respect to the heme plane during the rebinding of NO. We performed transient absorption measurements of the charge transfer Band III of human hemoglobin and raw transient spectra are displayed with kinetics at three different wavelengths as shown in Fig. 6. Apparently the kinetics at band III maximum (760 nm) is slower than that of the background (tail of the Q-bands: 760 nm and 780 nm).

The kinetics at 760 nm was fit to a multiexponential function with a low base line. It contains a fast (18 ps, 61 %) and a slow (140 ps, 24 %) component together with a constant. The time evolution at Band III was compared with that of kinetics of NO rebinding at Soret. The NO recombination components can be found as a fast process (11 ps, 74 %) and a slow process (61 ps, 22 %) together with a constant (4%) after analyzing the kinetics shown as Fig. 7.

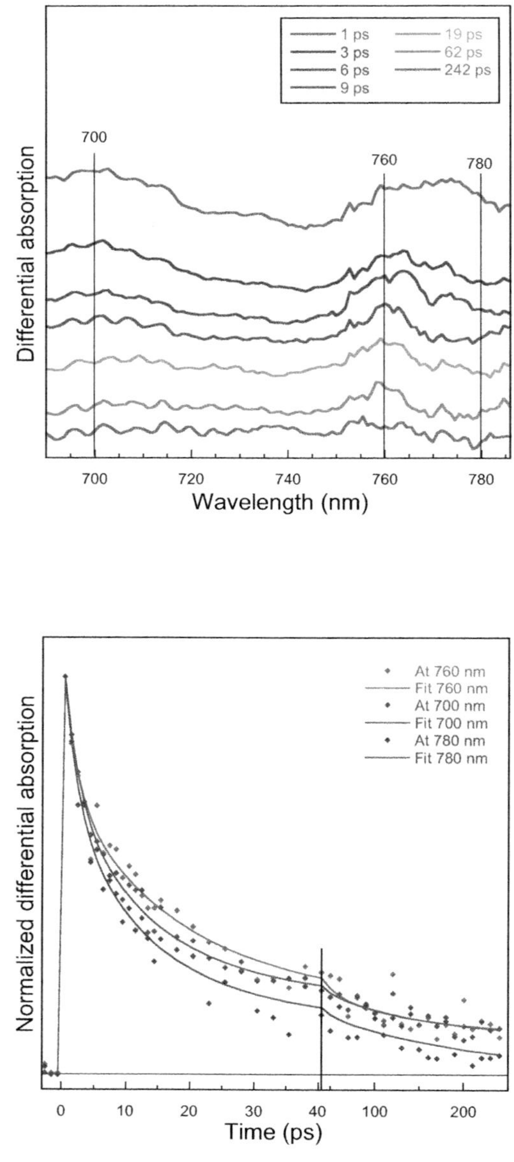

Fig. 6 Kinetics of the absorption Band III (755-760 nm, charge transfer transition) after Hb-NO photolysis, recorded with the same apparatus as the Soret absorption kinetics. Upper panel: raw transient spectra at selected time delays. Down panel: normalized kinetics at selected wavelengths, indicated by the dotted lines in upper panel.

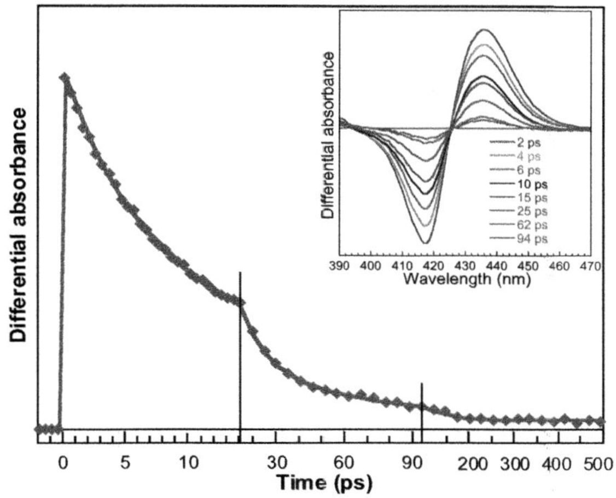

Fig. 7 Kinetics of NO rebinding to the human hemoglobin from SVD analysis of the entire time-wavelength matrix and raw transient spectra probed in Soret band (inset) at different time delays up to 500 ps

The dominant 18 ps decay component at Band III is clearly slower than that at Soret. Kinetics (11ps) represents the coordination of the heme after photodissociation while the Band III represents the motion of the iron (18 ps). This result is in agreement with the theoretical study on Band III intensity [20]. Because this latter kinetics is slower, we can deduce that the iron does not move instantaneously upon NO binding due to the protein relaxation which influences on this motion. This iron motion back into the heme plane represents the first step of the allosteric transition T→R upon NO binding.

5 Conclusion

Through the study of these two proteins, we have illustrated that time-resolved spectroscopy allows to identify the very first structural events which induce allosteric transitions in heme binding sites and consequently the activation and deactivation of the proteins. Importantly, to trigger these transitions, we have used the light absorption properties of the heme which allows one to cleave the chemical bond between the heme site and the diatomic ligand, synchronizing the population of molecules to be observed. Such a methodology has also been applied to non-naturally absorbing proteins such as photosystems, rhodopsin and fluorescent proteins [21, 22]. Now, time-resolved spectroscopy is awaiting its adaptation to unnaturally absorbing proteins, for example by using photo-activatable effectors or by introducing light-sensitive constituents in the protein by bio-engineering.

References

[1] Lancaster, J.: Nitric Oxide, Principles and Actions. Academic Press, San Diego (1996)
[2] Gautier, C., et al.: J. Biol. Chem. 279, 435 (2004)
[3] Negrerie, M., et al.: J. Biol. Chem. 276, 46815 (2001)
[4] Beal, D., et al.: Rev. Sci. Instrum. 70, 202 (1999)
[5] Kim, S., et al.: J. Phys. Chem. B108, 20366 (2004)
[6] Dartigalongue, T., et al.: Chem. Phys. Letters 415, 313 (2005)
[7] Kruglik, S.G., et al.: J. Raman Spectrosc. (2010) doi:10.1002/jrs.2685
[8] Kruglik, S.G., et al.: J. Biol. Chem. 282, 5053 (2007)
[9] Negrerie, M., et al.: J. Biol. Chem. 276, 46815 (2001)
[10] Negrerie, M., et al.: J. Phys. Chem. B 110, 12766 (2006)
[11] Bredt, D.S., et al.: Annu. Rev. Biochem. 63, 175 (1994)
[12] Balligand, J.L., et al.: Medecine & Science 15, 310 (1999)
[13] Schonhoff, C.M., et al.: J. Biol. Chem. 278, 18265 (2003)
[14] Fraser, M., et al.: Oncogene 25, 2203 (2006)
[15] Blaise, G.A., et al.: Toxicology 208, 177 (2005)
[16] Evgenov, O.V., et al.: Nature Reviews Drug Discovery 5, 755 (2006)
[17] Stone, J.R., et al.: Biochemistry 33, 5636 (1996)
[18] Pal, B., et al.: J. Raman Spectrosc. (2010); doi: 10.1002/jrs.2578
[19] Franzen, S., et al.: Nat. Struct. Biol. 1, 230 (1994)
[20] Stavrov, S.S.: Chemical Physics 271, 145 (2001)
[21] Kukura, P., et al.: Science 310, 1006 (2005)
[22] Aubert, C., et al.: Nature 405, 586 (2000)

Modeling an Active Peptide Conformation and Design a Competitive Inhibitory Peptide for HMG-CoA Reductase

Valeriy V. Pak[1,*], Vladimir N. Shin[2], and Lyubov Yun[1]

[1] Institute of the Chemistry of Plant Substances, Tashkent, Uzbekistan
pakvaleriy@yahoo.com
[2] Uzbekistan State Institute of Physical Culture, Tashkent, Uzbekistan

Abstract. The current work presents an approach that can be used to search for lead peptide candidates, including unconstrained structures in a recognized sequence. This approach was performed using the design of a competitive inhibitor for HMG-CoA reductase (HMGR). In a previous design for constrained peptides, a head-to-tail peptide cycle was used as a model of linear analogy in searches for lead peptides with a structure close to an active conformation. Taking into consideration the restricted flexibility of linear peptides in the binding site it was proposed that an analogical approach can be applied for finding a lead peptide with an unconstrained structure in a recognized sequence via modeling a cycle using fixed residues of the peptide backbone. The eight peptide cycles were selected for a peptide library based on the YVAE sequence as a recognized motif. For each cycle, the four models were assessed according to the design criterion (V) applied for constrained peptides. Three peptide cycles (FGYVAE, FPYVAE and FFYVAE) were selected as lead cycles from the library. The linear FGYVAE peptide (IC_{50} = 0.4 μM) showed a 1200-fold increase the inhibitory activity compared to the first isolated LPYP peptide (IC_{50} = 484 μM) from soybean.

1 Introduction

3-Hydroxy-3-methylglutaryl CoA reductase (HMGR) is the major regulatory enzyme of cholesterol biosynthesis and therefore constitutes the target enzyme of many investigations aimed at lowering the rate of cholesterol biosynthesis [1]. The efficacy of cholesterol biosynthesis inhibitors for the control of blood cholesterol levels is now well recognized. Their importance is linked with reducing the risk of hypercholesterolemia, which is known as a major risk factor associated with various cardiovascular diseases [2].

In previous works, the two hypocholesterolemic peptides (LPYP and IAVPGEVA) were found by analyses of a digested soy glycinin using trypsin and pepsin (respectively) [3,4]. An alignment of the amino acid composition of soy 11S-globulin with the IAVPGEVA sequence revealed another IAVPTGVA sequence with inhibitory activity against HMGR [5]. A conformational analysis of

* Corresponding author.

M.-W. Han and J. Lee (Eds.): EKC 2010, SPPHY 138, pp. 135–143.
springerlink.com

those peptides confirmed that a "turn" structure, which includes P-residue as a conformational constraint in the recognized motif, is a bioactive conformation of these peptides [6].

The design of a competitive inhibitory peptide for HMGR with a constrained structure based on the recognized VPTG sequence has previously been described [7]. A principle component analysis (PCA), which is projected multidimensional data on low-dimensional subspace, was applied to evaluate a head-to-tail peptide cycle as a model of linear analogy in order to select a lead peptide candidate. According to the conformational aspect, the binding affinity is defined as the percentage of overlap between a predefined set of conformations that can fit the binding site and the actual set of conformations that the molecule can adopt [8]. Based on the dependence of the binding affinity from the percentage of conformations that can fit into the binding site, and based on an analysis of the "region of bioactivity", as prescribed by the host molecule, it was proposed that fixed peptide conformations by active/recognized residues can be seen as an approach to model the restricted flexibility of the peptides in the binding site. Taking these considerations into account, a cyclic peptide with fixed residues was used as a model of a linear unconstrained analogy in searching for a less flexible sequence.

In the proposed design, the YVAE peptide was used as a basis of the recognized sequence, which presents an unconstrained peptide structure. It was selected in accordance with the strongest ability to inhibit HMGR in a competitive manner among previously designed peptides [9]. A two-stage approach was applied in the design of the peptide. The first was the evaluation of an additional space to extend the peptide length. For the YVAE peptide, the designed space was determined by superposition of the bioactive conformation of simvastatin and rosuvastatin, which partially includes the space occupied by the central part of the statins. A structure-functional analysis of the statins suggests that these rigid hydrophobic groups play an important role in binding [10]. From this point of view, the structures of the statin molecules were investigated in accordance with the structural diversity of their central parts. Through an analysis of the statin molecules belonging to the two types structures due to the structural diversity observed in the rigid hydrophobic groups, atorvastatin, fluvastatin and simvastatin were selected as model compounds for assessing the space occupied by the peptides [11]. The second stage was an assessment of the conformational stability of the peptide models using the design criterion ("V" parameter) applied for constrained peptides [7]. The biological activity of the designed peptides was then assayed through the use of an *in vitro* test.

2 Materials and Methods

2.1 Materials

H-Glu(OtBu)-2-ClTrt resin (substituted at 0.55 meq/g) and Fmoc-amino acids were purchased from AnaSpec (San Jose, CA, USA). Chemicals for the peptide synthesis were obtained from Perkin Elmer (Foster, CA, USA). Acetonitrile and methanol for HPLC were the products of Burdict & Jackson (Muskegon, MI, USA).

2.2 Peptide Preparation

Syntheses of the peptides were performed using the standard fluorenylmethyloxycarbonyl (Fmoc) methodology with an automated synthesizer process utilizing an Applied Biosystem Peptide Synthesizer (Model 433A, Perkin Elmer, Foster, CA, USA). Peptide synthesis starting with the E-preloaded resin was carried out using a standard FastMoc strategy supplied by Perkin Elmer [12]. The peptides were cleaved from the resin via a mild trifluoroacetic acid cleavage method [13].

Purification and analyses of the synthetic peptides were done with small variations of the procedures described in a previous study [9]. Peptides were purified using a reversed phase-high performance liquid chromatograph (RP-HPLC) (Waters, Milford, MA, USA) and were identified using an ESI-MS (Platform II, Micromass, Manchester, UK) and an Applied Biosystems 491 Peptide Sequencer (Perkin Elmer, Foster, CA, USA).

2.3 Assay of HMG-CoA Reductase Inhibitory Activity

The HMG-CoA-dependent oxidation of NADPH, used for assessing the HMGR activity, was monitored at 340 nm on a Jasco V-530 spectrophotometer (Model TUDC 12B4, Tokyo, Japan). Assay conditions were used as described in the aforementioned earlier study [9]. One unit (U) of HMGR was defined as the amount of enzyme that catalyzes the oxidation of 1 μmol of NADPH per min. The protein concentration was determined using the method of Bradford [14]. The type of inhibition and Michaelis-Menten parameters were determined from a Dixon plot [15].

2.4 Computational Methods

The structures of the peptides were built using the program package ChemOffice Desktop 2004 for Windows (CambridgeSoft (CS) Corporation, MA, USA). The structures of the cyclic peptides used in the design of the peptide sequences were carried out using a molecular mechanics method (MM2) within Chem3D [16]. Calculations of the optimized structure for peptide models were carried out using the AM1 method within the CS MOPAC (Version 1.11) program package [17]. Molecular dynamic (MD) simulation was used for the estimation of the peptide backbone behavior. This was performed using ChemOffice Desktop 2004 for Windows. The peptide structures were collected in 20 ns and were heated to 300 K.

The design criterion "V" for the peptide models was calculated as described in a previous study [7]. Geometry-optimized low-energy and MD conformers of models of the cyclic peptides were collected using a slightly modified method of that in study [18].

3 Results and Discussion

3.1 Peptide Design

In order to design a linear peptide with an unconstrained structure in a recognized sequence, two subsequent stages were applied. As a starting point for the first stage, structures of all statin molecules extracted from the crystal structure of HMGR-statin complexes were analyzed while focusing on the structural diversity of the rigid hydrophobic groups.

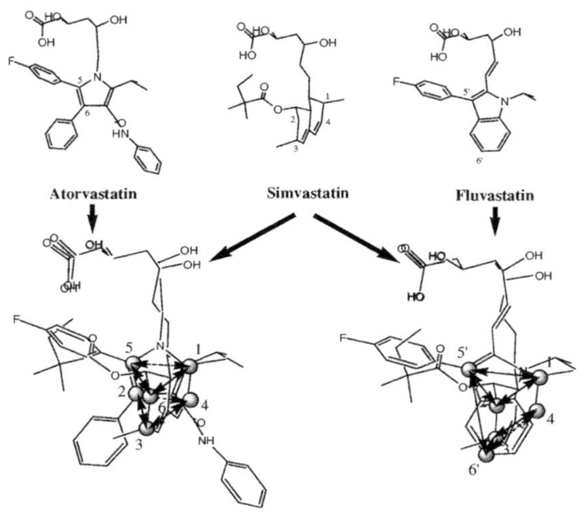

Fig. 1 Modeling of a designed space using the superposition of the bioactive conformations of atorvastatin and fluvastatin with simvastatin extracted from the crystal structure of HMGR-statin complexes (PDB codes: 1HWK (atorvastatin), 1HWI (fluvastatin), 1HW9 (simvastatin)). The finite space defined by atoms 1-2-3-4-5'-6' was used as a model of the designed space in the present study.

A structure-functional analysis of these statin parts led to the proposition that they carry out the directing function for the attached radicals. This implies modeling of these parts using a relatively rigid structure of a peptide backbone with side-chains, conforming to the hydrophobic radicals of statins. By a superposition of the bioactive conformations of atorvastatin, fluvastatin and simvastatin (PDB codes: 1HWK (atorvastatin), 1HWI (fluvastatin), 1HW9 (simvastatin)), the available space for modeling was assessed. The superposition of the bioactive conformations was constructed by considering that the HMG-moiety of statins is bound in the narrow pocket of the active site, which is a common feature in statin-binding by HMGR. As shown in Figure 1, the two contrastingly oriented planes pass through the decalin ring of simvastatin and the pyrrolyl ring of atorvastatin, as well as through the indol ring of fluvastatin. These findings led to the proposition that finite spaces defined by atoms 1-2-3-4-5-6 for the superposition of atorvastatin with simvastatin and by atoms 1-2-3-4-5'-6' for the superposition of fluvastatin with simvastatin can be used to assess the number of added amino acids through a comparison of their occupied volumes.

To select the length of the added amino acids, the measured means of the defined spaces using statin molecules were compared to those of the volumes occupied by the peptide backbones with one, two and three amino acid residues. Using the geometry of the peptide structures obtained through molecular dynamic (MD) studies, it was found that the ratios of the occupied volume by the peptide backbones using one, two and three residues, to the space determined for the atorvastatin-simvastatin superposition (1-2-3-4-5-6) were equal to 1.83±0.23, 3.42±0.31 and 5.32±0.28, respectively. This finding suggests that this space presents a small volume for simulation, even if using one additional residue. For space 1-2-3-4-5'-6', obtained through the superposition of fluvastatin and simvastatin, these values were 0.35±0.18, 0.67±0.26 and 1.04±0.22, respectively. The consistency of these results suggests that a peptide length of two amino acids is required for optimal simulation of the central part of the statins.

Thus, the whole length of the peptide was selected as six amino acids with E-residue in the C-terminus. By considering the hydrophobic radicals included in rigid hydrophobic groups of atorvastatin and fluvastatin, the peptides FGYVAE, FAYVAE, FPYVAE, FVYVAE, FLYVAE, FIYVAE, FFYVAE and FWYVAE were chosen as candidate peptides for the peptide library.

The second stage was to investigate the conformational stability of the peptide models. A head-to-tail cyclization was used for each peptide sequence from the library. Four sets were defined for models of a cyclic peptide based on various numbers of the fixed recognized residues. The sets of the peptide models were selected as follows: Set 1 included the peptide models without fixed atoms of the peptide backbone; Sets 2, 3 and 4 were peptide models in which the atoms of the peptide backbone were fixed at two, three and four residues from the C-terminus, respectively. According to a previous design for the YVAE peptide, the E-residue is a mimic of the HMG moiety [9]. Taking this into account, atoms of the side-chain of the E-residue were fixed in all sets of peptide models.

The dihedral angles of the active peptide backbone for the YVAE peptide, defined in an earlier study, were used in the construction of the first peptide model in each set [19]. To evaluate the effect of the fixed residues on the conformational behavior of the peptide cycle, 100 low-energy conformers in each set were additionally studied by MD simulation. Figure 2 demonstrates a similar pattern in the alterations of the "V" parameter for the conformers that were investigated in each set. This finding shows that determination of a less flexible cycle stems from a fixed number of residues and does not depend on the starting structure of peptide cycle. The consistency of these results suggests that the proposed approach can be applied in the design the linear peptides, which include both unconstrained and constrained structures in a recognized sequence.

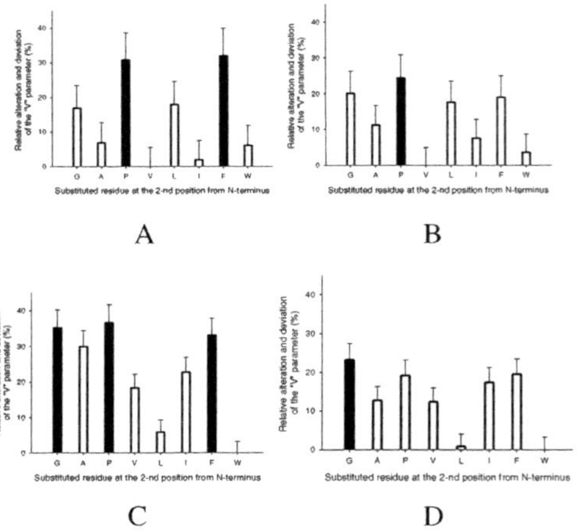

Fig. 2 The dependence of the relative alteration of the "V" parameter for the peptide models in each set. (A) Set 1: Atoms of the side-chain of the E-residue were fixed. (B) Set 2: Atoms of the peptide backbone of the EA fragment were fixed. (C) Set 3: Atoms of the peptide backbone of the EAV fragment were fixed. (D) Set 4: Atoms of the peptide backbone of the EAVY fragment were fixed. Atoms of the side-chain of the E-residue were fixed in all sets of the peptide models. The resulting means in percentages were estimated as the relative alterations of peptide stability compared to the most flexible peptide cycle in each set. The value "0" corresponds to the FVYVAE and FWYVAE sequences in sets "1", "2" and "3", "4", respectively. Peptides marked in black were proposed as active compounds.

The most rigid structures in Set 1 were obtained for the FPYVAE and FFYVAE sequences. The estimated data of the peptide models in Set 2 revealed the FPYVAE sequence as a less flexible cycle. In Set 3, the most rigid models were determined for the FGYVAE, FPYVAE and FFYVAE sequences, and the FGYVAE sequence was the lead cycle in Set 4. A comparison the values of the relative alteration from the most flexible structure in all sets revealed that the highest means were obtained in Set 3 (Figure 2C). These findings propose that the FGYVAE, FPYVAE and FFYVAE sequences can be considered lead peptide candidates.

Thus, these peptides were synthesized and their biological activities were estimated. The IAVPGEVA and IAVPTGVA peptides were selected as control compounds in this study. According to Set 3, in which all three sequences were determined as lead peptide candidates, the FWYVAE peptide was chosen to assess the proposed approach as the most flexible peptide in this set. The linear GGYVAE and GGGGYVAE peptides were additionally synthesized to evaluate the design criterion ("V" parameter) in accordance with the previous study [7]. In the FGYVAA peptide, the recognized E-residue was substituted with the A-residue to provide a negative control for a binding study.

3.2 Inhibition Activities of Designed Peptides

An assay of the *in vitro* susceptibility of HMGR to inhibition by the synthetic peptides was performed in a range of peptide concentrations from 0.2 to 1000 μM. Each of the synthesized peptides showed some ability to inhibit HMGR, with the exception of the FGYVAA peptide. No activity was detectable when an upper bound of the concentration range was used (Table 1). This finding supports previously obtained data that showed that the active site of HMGR recognizes the E-residue as a recognized component for the HMG-biding pocket.

Table 1 Peptide sequences, peptide models and inhibitory activities (IC_{50}) of the synthesized peptides used in this study

Peptide	Peptide sequence	Fixed residues in set				IC_{50} [μM]
		1	2	3	4	
Design	GGYVAE	E	EA	EAV	EAVY	7.4
	FFYVAE	E	EA	EAV	EAVY	2.5
	FPYVAE	E	EA	EAV	EAVY	1.4
	FGYVAE	E	EA	EAV	EAVY	0.4
	FWYVAE	E	EA	EAV	EAVY	29.5
Control	GGGGYVAE	E	EA	EAV	EAVY	760.7
	IAVPTGVA	T	TG	TGV	TGVA	152.1
	IAVPGEVA	E	EV	EVA	EVAI	201.3
	LPYP[a]					484.7
	YVAE[b]					41.9
Negative control	FGYVAA					NA

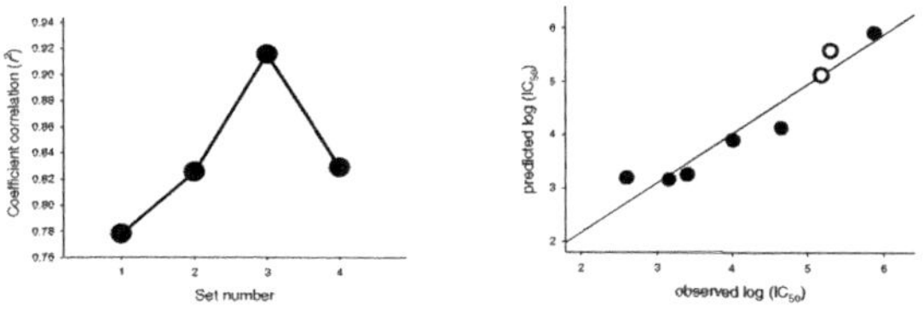

A B

Fig. 3 The correlation between the observed peptide potency and the predicted potency against HMGR. (A) The determined correlation coefficients in each set. (B) Designed peptides and control peptides (IAVPTGVA and IAVPGEVA) measured *in vitro* assay for Set 3. The correlation coefficient, r^2, is 0.91. The designed peptides are marked as (•); the control peptides are marked as (○).

A significant increase in the inhibitory activities was found for the linear FPY-VAE, FGYVAE and FFYVAE peptides, which were selected as lead compounds. A six-fold increase of the inhibitory activity for the GGYVAE compared to the YVAE peptide indicates the contribution of the peptide backbone of the two G-residues during binding. As was expected, the inhibitory activity of the FWYVAE peptide was found to be low compared to the lead peptide candidates in accordance with the value of the "V" parameter.

To estimate the proposed approach, the dependence between the observed and predicted peptide potency was plotted using the values of the "V" parameters for the peptides in each set. Figure 3A shows the determined correlation coefficients in the sets. The highest value of the correlation coefficient obtained in Set 3 indicates the best model for this library. In this set, a relatively good correlation (r^2 = 0.91) was determined between the experimental bioactivity of the linear peptides, measured by log (IC_{50}) and the predicted bioactivity obtained through the "V" parameter (Figure 3B). The difference between the obtained values of the peptide bioactivities was checked with the help of the Fisher criterion. An insignificant difference between these values was determined using the Fisher criterion ($F_{exp} < F_{(0.05, f1, f2)}$). This finding suggests that the proposed approach is acceptable in the design of linear peptides.

To clarify the properties of HMGR inhibition by the designed peptides, the enzyme activity was measured with HMGR in the absence and in the presence of FGYVAE, as this was the most potent inhibitor in this study. A Dixon plot shows that FGYVAE inhibited the HGMR in a competitive manner (Figure 4). The equilibrium constant of inhibitor binding (K_i) for the inhibition of HMGR by FGY-VAE was estimated to be 0.14±0.01 μM, which is slightly over 100 times lower compared to YVAE (15.2±1.4 μM).

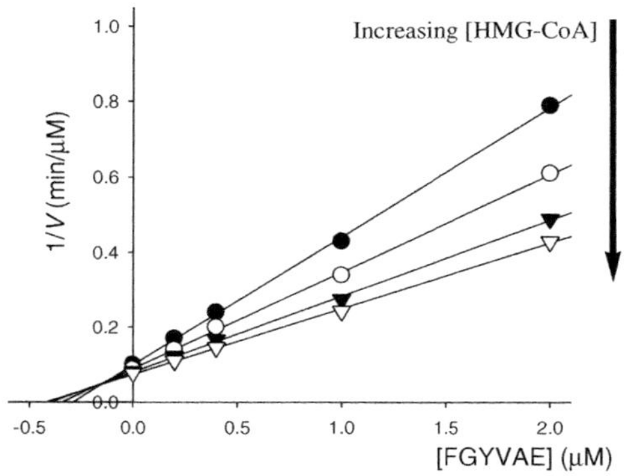

Fig. 4 The kinetics of inhibition of HMGR by the FGYVAE peptide. The data are presented in a Dixon plot, $1/V$ against [FGYVAE], at a NADPH concentration of 120 μM and HMG-CoA concentrations of 72 μM (•), 96 μM (○), 120 μM (▼) and 144 μM (□). The K_M value was found to be 68±3.2 μM for HMG-CoA and the V_{Max} value was 19.4±0.4 μM/min. Each straight line represents the results of triplicate experiments.

4 Conclusion

The proposed method presents an approach to design an active peptide backbone for an HMGR inhibitory peptide. The peptide models with fixed residues based on the design of a suitable library of flexible molecules allows the examination of the conformational behavior of members of the library in accordance with the recognized features of the peptides. The assumed correlation between the conformational behaviors of the models of the cyclic peptides obtained by MD simulation and the restricted flexibility of the peptides in the binding site is used when searching the lead compounds. In this aspect, a recognized sequence, which includes a constrained structure, can be seen as a special case of the presented approach. The obtained space occupied by the active peptide conformations implies the design of a more tightly bound peptide by the HMGR active site. Modeling an active backbone for peptides with different peptide lengths and substituted residues is a good opportunity to create an active peptide with adjusted properties that may be applied to ligands in investigation of peptide-protein interactions.

Thus, the present study not only shows the design of a more potent inhibitor for HMGR, but also defines a design tool to model active conformations for linear peptides.

References

[1] Endo, A.: J. Lipid Res. 33, 1569 (1992)
[2] Eisenberg, D.: Am. J. Med. 104, 2S (1998)
[3] Kwon, D., et al.: Food Sci. Biotechnol. 11, 55 (2002)
[4] Pak, V., et al.: Chem. Nat. Comp. 41, 710 (2005)
[5] Pak, V., et al.: Chem. Nat. Comp. 41, 454 (2005)
[6] Pak, V., et al.: Chem. Nat. Comp. 40, 398 (2004)
[7] Pak, V., et al.: J. Mol. Recogn. 20, 197 (2007)
[8] Becker, O.: Proteins 27, 213 (1997)
[9] Pak, V., et al.: Biopolymers. Pept. Sci. 84, 586 (2006)
[10] Istvan, E., et al.: Science 292, 1160 (2001)
[11] Istvan, E.: E. Atherosclerosis suppl. 4, 3 (2003)
[12] Fields, C., et al.: Pept. Res. 4, 95 (1991)
[13] Elmer, P.: Introduction to Cleavage Techniques, 9 (1995)
[14] Bradford, M.: Anal. Biochem. 72, 248 (1976)
[15] Segel, I.: Biochemical Calculations, 246 (1976)
[16] Burkert, U., et al.: Molecular Mechanics (1982)
[17] Dewar, M., et al.: J. Am. Chem. Soc. 107, 3902 (1985)
[18] Kalazsi, A., et al.: J. Peptide Sci. 8, S194 (2002)
[19] Pak, V., et al.: Bioorg. Med. Chem. 16, 1309 (2008)

Author Index

GPSR Compliance

The European Union's (EU) General Product Safety Regulation (GPSR) is a set of rules that requires consumer products to be safe and our obligations to ensure this.

If you have any concerns about our products, you can contact us on ProductSafety@springernature.com

In case Publisher is established outside the EU, the EU authorized representative is:

Springer Nature Customer Service Center GmbH
Europaplatz 3
69115 Heidelberg, Germany

Batch number: 10755292

Printed by Printforce, the Netherlands